CREATION, EVOLUTION AND MEANING

This book presents the case for belief in both creation and evolution at the same time as rejecting creationism. Issues of meaning supply the context of inquiry; the book defends the meaningfulness of language about God, and also relates belief in both creation and evolution to the meaning of life. Meaning, it claims, can be found in consciously adopting the role of stewards of the planetary biosphere, and thus of the fruits of creation.

Distinctive features include a sustained case for a realist understanding of language about God; a contemporary defence of some of the arguments for belief in God and in creation; a sifting of different versions of Darwinism and their implications for religious belief; a Darwinian account of the relation of predation and other apparent evils to creation; a new presentation of the argument from the world's value to the purposiveness of evolution; and discussions of whether or not meaning itself evolves, and of religious and secular bases for belief in stewardship.

Transcending Boundaries in Philosophy and Theology

Series Editors:
Martin Warner, University of Warwick, UK
Kevin Vanhoozer, Trinity International University,USA

Transcending Boundaries in Philosophy and Theology is an interdisciplinary series exploring new opportunities in the dialogue between philosophy and theology that go beyond more traditional 'faith and reason' debates and take account of the contemporary reshaping of intellectual boundaries. For much of the modern era, the relation of philosophy and theology has been conceived in terms of antagonism or subordination, but recent intellectual developments hold out considerable potential for a renewed dialogue in which philosophy and theology have common cause for revisioning their respective identities, reconceiving their relationship, and combining their resources. This series explores constructively for the 21st century the resources available for engaging with those forms of enquiry, experience and sensibility that theology has historically sought to address. Drawing together new writing and research from leading international scholars in the field, this high profile research series will offer an important contribution to contemporary research across the interdisciplinary perspectives relating theology and philosophy.

Also in this series:

Kierkegaard, Language and the Reality of God
Steven Shakespeare

Impossible God
Derrida's Theology
Hugh Rayment-Pickard

On Paul Ricoeur
The Owl of Minerva
Richard Kearney

Creation, Evolution and Meaning

ROBIN ATTFIELD
Cardiff University, UK

ASHGATE

Published by
Ashgate Publishing Limited
Gower House
Croft Road
Aldershot
Hants GU11 3HR
England

Ashgate Publishing Company
Suite 420
101 Cherry Street
Burlington, VT 05401-4405
USA

Ashgate website: http://www.ashgate.com

British Library Cataloguing in Publication Data
Attfield, Robin
Creation, evolution and meaning. – (Transcending boundaries in philosophy and theology)
1.Creation 2.Evolution – Religious aspects 3.Evolution (Biology) – Religious aspects 4.Cosmology
I.Title
113

Library of Congress Cataloging-in-Publication Data
Attfield, Robin.
Creation, evolution and meaning / Robin Attfield.
p. cm. – (Transcending boundaries in philosophy and theology)
Includes bibliographical references and index.
ISBN 0-7546-0474-8 (hardcover: alk. paper)
1. Creation. 2. Evolution (Biology) 3. Creationism. 4. Meaning (Philosophy) 5. Stewardship, Christian. 6. Philosophical theology.
I. Title. II. Series.
BS651.A77 2006
213--dc22

2006008831

ISBN-13: 978-0-7546-0474-7; ISBN-10: 0-7546-0474-8 (Hbk)
ISBN-13: 978-0-7546-0475-4; ISBN-10: 0-7546-0475-6 (Pbk)

Typeset by Pat FitzGerald

Printed and bound in Great Britain by MPG Books Ltd, Bodmin, Cornwall.

Contents

Acknowledgements *ix*

Introduction 1

PART I: MEANING AND CREATION

Chapter One: Meaning, Verification and Analogy 7
Section 1: Meaningful Communication, a Meaningful Life and Theories of Meaning 7
Section 2: The Verificationist Challenge 9
Section 3: Some Implications of Verificationism for Religious Language 13
Section 4: Ayer's Subsequent Stances 16
Section 5: Some Objections to Verificationism 19
Section 6: Creation and Analogy 24

Chapter Two: Realism, Anti-realism and Religion 29
Section 1: Realism and the Roots of Anti-realism 29
Section 2: Dummett's Anti-realist Arguments 31
Section 3: Further Arguments for and against Anti-realism 36
Section 4: Rorty's Anti-realism 40
Section 5: Cupitt's Anti-realist Faith 42
Section 6: Phillips's Religious Projectionism 45

Chapter Three: God and Falsification 51
Section 1: Popper, Duhem and Quine 51
Section 2: The Falsification Controversy 53
Section 3: Creation and Falsification 56
Section 4: Falsification, Religion and Explanations 59
Section 5: Findings So Far about Creation 61

PART II: CREATION AND EVOLUTION

Chapter Four: Creation 67
Section 1: Some Alternative Conceptions 67
Section 2: Transcendence 69
Section 3: Creation, Necessity and Contingency 73

Section 4: Creation and Time 76
Section 5: Creation without Creationism? 80
Section 6: Darwin on Creation 84

Chapter Five: Arguments from World to God 89
Section 1: From World to God 89
Section 2: The Cosmological Argument 93
Section 3: Varieties of Design Argument 97
Section 4: Hume's Critique of the Design Argument 101
Section 5: Fine Tuning, Universes and Probability 106

Chapter Six: Darwinism, Disvalues and Design 109
Section 1: Darwinism 109
Section 2: Darwinism, Theism and Gould 115
Section 3: Darwinism, Theism and Ruse 118
Section 4: Disvalues in Nature 121
Section 5: Nature and Immorality 125

Chapter Seven: God and Evil 133
Section 1: Are Theism and Evil Compatible? 133
Section 2: Atrocities 136
Section 3: Natural Evil 139
Section 4: Dawkins on Predation 143
Section 5: Could Things be Otherwise? 147

Chapter Eight: Purpose, Immanence and the Argument from Value 151
Section 1: Ward's Value Argument 151
Section 2: Intrinsic Value 154
Section 3: The Argument from Intrinsic Value Revisited 158
Section 4: The Argument from Value to Progressivism Reconsidered 162
Section 5: Purpose, Immanence and Panentheism 166

PART III: EVOLUTION AND MEANING

Chapter Nine: Meaning, Evolution and Stewardship 175
Section 1: Introduction 175
Section 2: Meaning, Value and Meaningful Action 175
Section 3: The Evolution of Design, Intelligence and Meaning – Dennett's Tower 178
Section 4: Some Criticisms of Dennett's Tower 181
Section 5: Meaningful Action, Meaningful Work and a Meaningful Life 186
Section 6: Darwinism and Meaningful Life 189
Section 7: Stewardship 191

Chapter Ten: The Ethics and Metaphysics of Stewardship 197
Section 1: Introduction 197
Section 2: Ethics: Scope, Method and Content 197
Section 3: On the Metaphysics of Stewardship 201
Section 4: Givenness, Metaphysics and Stewardship 203
Section 5: Creation, Evolution and Meaning 206

Bibliography *211*
Index *229*

Acknowledgements

Thanks are in place to Cardiff University for granting research leave for the Spring semester of 2004–2005, to the Arts and Humanities Research Council for funding research leave for the Autumn semester of 2005–2006 to enable this book to be completed, and to Frederick Ferré for his generous perseverance in supporting applications for this funding. I would also like to thank colleagues for undertaking my teaching and administration during this twelve-month period, and to Barry Wilkins for master-minding all these arrangements. Special thanks are owed to Martin Warner, Stephen Moller, David Attfield and Leela Attfield for reading and commenting on a draft of the whole book, and to David Edwards, Christopher Norris and Arthur Peacocke for commenting on particular chapters. I am also grateful to Andrew Belsey and to Alexander Miller for valuable ideas and suggestions. All these people have enhanced the outcome, although probably none of them would endorse it in every particular; certainly none of them carries responsibility for such shortcomings as remain. Above all, I am particularly grateful to my wife Leela for constant support and encouragement, without which this book could not have been written.

Introduction

Belief in creation is often misleadingly pitted against Darwin's theory of evolution, and is prone to be confused with creationism, which conflicts with that theory. One of the main arguments of this book is that there is no conflict between evolution and belief in creation, and that belief in creation need not involve adherence to creationism, either in its Victorian version, or in its contemporary, consciously anti-Darwinian form, or even when dressed up in the trappings of Intelligent Design. Intelligent Design theorists invoke supernatural interventions to explain (what Intelligent Design considers to be) the phenomena of 'irreducible complexity'; they give up on finding naturalistic explanations for these phenomena, and yield to premature despair instead. Theistic evolutionists, by contrast, do not reject belief in a designing intelligence, but can consistently recognise a creator, whose creation is completed through naturalistic processes such as natural selection – or so this book contends.

In order to present this case, a wide-ranging overview of varieties of Darwinism is introduced, but not until what is meant by belief in creation has been clarified. Creation, it is contended here, concerns not the Big Bang or some other earliest event, but the dependence of each and every physical entity on a divine creator, not situated in space or in time, possessed of the power and the knowledge to bring the world into being, and to select its natural laws. After the meaning of belief in creation has been clarified, grounds are supplied for this being a rational belief, grounds which are shown not to have been overthrown by Darwinism, let alone by philosophical critics. Darwinism in effect presents one of the laws of nature for which creation supplies the best explanation. The case against creation (consisting in the world's evils) is also considered, both in its pre-Darwinian and its Darwinian forms (and this is where the varieties of Darwinism become important). It is here argued that no irregular world (such as a world of miraculous interventions) would be preferable to the actual law-governed world, and also that no other world with different laws can be known to be more valuable either.

However, these issues cannot be considered until prior issues of coherence and of meaning have first been addressed. This book in fact concerns meaning in two senses, the meaning of language (in Part I) and the meaning to be found in meaningful action or a meaningful life (in Part III). (It is later suggested that our use of the same word in both connections is no coincidence.) The theory that theistic language (language about God) is meaningless (now unfashionable but far from defunct) is first investigated, but rejected; and then the analogical nature of most theistic language is introduced (a key theme in view of the debates about such language that come into prominence later). Next, the anti-realist theory that we cannot grasp language that transcends possible verification, nor employ about such language terms like 'true' in their ordinary sense, is addressed, initially over its philosophical grounds, and subsequently in connection with specific applications to God and creation. But these theories of meaning turn out to be unsubstantiated, both in their generic form (applying to all our language) and also in specific versions relating to religion. The further theory that there can be no genuine statements that are not falsifiable is more favourably treated, but shown to require qualification; once qualified, the criterion of falsification turns out actually to be satisfied by belief in creation, given a principle governing explanations that is here introduced. Indeed it is creation that makes experience possible.

The value to be found in the world is further discussed towards the end of Part II ('Creation and Evolution'). Here, in Chapter Eight, the notion of non-derivative or intrinsic value is first clarified, and then argued to characterize the world's creatures enough to warrant belief in a creator desirous of value, and whose creation generates value through evolutionary processes. Here an argument initiated by Keith Ward, but found to be defective, is overhauled and argued in its revised form to support these conclusions. It is concluded that there is a purpose underlying evolution, not expressed in the kind of interventions alleged by the theorists of Intelligent Design, but embodied, for example, in the selection of laws of nature finely tuned for the possibility of life, and in conditions facilitating its evolution on planets such as our own. Contrary to theorists who suggest that either complexity or consciousness form the goal or purpose of evolution, value is located in the flourishing of every life-form, while greater value is detected in the development of capacities such as the forming of conscious relationships.

This sets the scene for reflection (in Part III) on meaning in the sense of the meaning of action and of life. The suggestion of equating meaning and value is rejected, and the theory that meaning (in all senses) evolves through Darwinian processes is heavily qualified, since meaning is largely a feature of human culture and generated by ordinary human (as opposed to natural) selection. Life's meaningfulness turns out to involve an integrated sense of priorities, self-awareness and a sense of objective values which the people concerned can see themselves as safeguarding or honouring or promoting. In view of the value of the products of evolution, this can take the form of understanding oneself as a steward or trustee of such value. Those who see themselves (or who see humanity) in this light can do so on either a theistic or a secular basis. Secular stewardship can be recognized as involving answerability to humanity or to the moral community, and can thus supply a basis for life's meaningfulness; but theistic stewardship (motivated by answerability to God the creator, regarded as the source of the world's value) has a greater coherence, and much more directly makes life meaningful.

In general, none of the variants of Darwinism need be regarded as undermining for its adherents the possibility of meaningful lives, despite the dangers of some forms that appear to treat as the real agents of history our genes, plus gene-like viruses, or 'memes', that supposedly infect minds with ideas, and appear to treat human lives as their mere carriers or containers. However, in non-deterministic versions that allow individuals to resist their genes, and particularly in versions that recognize ideas as generated not by memes but in some degree by rational choices, even the kind of adaptationist Darwinism that is closest to these theories allows people to see themselves as privileged to be guardians or custodians of nature's value. This being so, the other variants of Darwinism, including the moderate materialist form (which restricts Darwinism to biological as opposed to cultural phenomena), and Mind-First Darwinism, which recognizes natural selection as an expression of a divine purpose, are all the more obviously conducive to an awareness of life's meaningfulness and the world's value.

Mind-First Darwinism also sees stewardship of the world's value as work entrusted to humanity by the creator of the value generated in the course of evolution. It is the central claim of this book that this view is both coherent, consistent and well-grounded.

PART I
Meaning and Creation

Chapter One

Meaning, Verification and Analogy

Section 1: Meaningful Communication, a Meaningful Life and Theories of Meaning

Some philosophers of meaning, such as Michael Dummett, claim that the theory of meaning (the meaning of sentences, that is) is the most fundamental sector of philosophy, while others question this. Others again hold that the study of meaningfulness depends for its value on throwing light on the meaning of life, and on the links between meaningful action (including meaningful communication) and the shared human practices that seem to give life its meaning. Both these concerns figure in this book, the first of them in the opening chapters. But before that discussion begins, it is worth pausing to reflect on the links between the meaning to be found in language and the meaning to be found in life.

When maintaining that 'the theory of meaning ... is the foundation for all philosophy', Dummett says by way of explanation that the first task of philosophy is 'the analysis of meanings', and that 'the deeper such analysis goes, the more it is dependent on a general account of meaning' and 'a model for what the understanding of an expression consists in'.[1] To some of the details of his claims we shall be returning in the next chapter; but these remarks already convey the importance of theories of meaning for most areas of thought, including reflection about creation. On the other hand, if the role of philosophy includes arriving at a defensible world-view, including perhaps a metaphysic and an ethic, then a semantic theory will comprise no more than one component of what it seeks.[2] And if a defensible world-view is to incorporate an account of what makes life meaningful, then a philosophy focussing on the meaning of language may leave the central task of philosophy not only unachieved but actually unattempted.

Is it, then, a mere coincidence that we speak both of the meaning of expressions and of the meaning of life? The obvious bridge concept is that of meaningful action, of which meaningful communication is a paradigm instance, since meaningful action typically plays a large role in a meaningful life. Both meaningful action and a meaningful life (topics to which we return in Part III) may be held to involve the holding of a set of more or less integrated priorities.

---

1 Michael Dummett (1973), *Frege: Philosophy of Language*, p. 669.

2 Alexander Miller (forthcoming), 'Realism and Anti-Realism', in E. Lepore and B. Smith (eds), *Oxford Handbook on the Philosophy of Language*, Section 4: 'The Plausibility of a Realistic Worldview'.

And meaningful communication may be held to derive its significance, in part, against this background. Simultaneously it allows significant transactions to take place, and expressions of the shared concepts of one or another culture, or of humanity, to be exchanged. Hence the very expressions are frequently brimming over with a matching significance, a significance not diminished but facilitated by their being encoded in language. All this suggests that our speaking both of the meaning of expressions and of the meaning of life is no coincidence.

Yet linguistic meanings belong to expressions that are banal or trivial as well as to those that shake our foundations or reinvigorate our sense of life's meaning. Correspondingly theories of meaning, while capable in principle of relating to and even of illuminating issues of private, public or even cosmic significance, may quite properly carry a significance restricted to the semantic field, seeking to tell us what makes communication and understanding possible, whether that communication is itself inspiring or insipid. For such theorists to bring off this aspiration would apparently be a worthwhile achievement even if there were no implications for understanding the riddles of the meaning of life. Just as linguistic meaning is sometimes unrelated to the meaning of life, so theories of meaning may have no aspirations to explain meaning in that sense.

However, philosophers of meaning of several stripes have claimed that theories of meaning are fundamental because they specify the nature and limits of what can be communicated and of what can be understood. If, as pragmatists claim, only statements with practical consequences have significance, then the consequences of their theory for what we can understand are themselves significant. If, as verificationists claim, language that cannot empirically be verified is literally meaningless, except when it derives its meaning from the senses of the terms employed, then great swathes of language lie (and have always lain) beyond our understanding (talk about creation and some talk about the meaning of life included). And if, as anti-realists claim, language concerning whatever transcends verification cannot be understood, then whatever might be known and understood by beings of ampler powers, the scope and content of human knowledge and understanding are significantly restricted, and the very notion of truth needs to be reinterpreted accordingly.

Accordingly it is important to pay some attention to such theories of meaning, as I seek to do in this chapter and the next two. For the topics of some later chapters, such as belief in creation and the meaning of life, would need to be differently interpreted if these theories are true, if not discarded altogether. The belief held in some quarters that such theories of meaning are no longer defended turns out to be illusory, since there are many defenders of pragmatism and of anti-realism, while verificationism continues to be held in updated forms.[3] Indeed anti-realism is sometimes held by religious believers who reject belief in a transcendent creator and interpret religious language accordingly;[4] hence this

3 See C.J. Misak (1995), *Verificationism: Its History and Prospects.*

4 See Peter Byrne (2003), *God and Realism*, p. 4.

form of anti-realism will also be discussed (Chapter Two). So too will be the theory that only claims capable of being falsified are meaningful as statements (Chapter Three). Discussion of these theories of meaning may at the same time serve to throw some light on meaning itself, and on the meaning of religious language in particular.

Section 2: The Verificationist Challenge

Before realism and anti-realism can be considered (see Chapter Two below), the challenge of verificationism to religious language needs to be discussed, not least because (as is explained in this section) verificationism poses no challenge to anti-realist interpretations of religion that represent religion as (for example) ethics in disguise, but presented an apparently formidable challenge to the kind of realist interpretations endorsed by most believers. Indeed, leading philosophers continue to debate the merits and demerits of the verificationism of the positivists.[5] It also proves valuable to consider the parallel but divergent verificationism of pragmatists such as C.S. Peirce, which arguably turns out to make room for reasoning to the best explanation, even if this could lead to the conclusion of belief in divine creation in explanation of sensory experience and its objects (see Section 3 below). The recent incisive critique of Abigail L. Rosenthal of the near-to-death experience of A.J. Ayer discloses significant limits to the scope of his later verificationism, thus throwing new light on the verificationist challenge (Section 4). But in any case verificationism proves to be subject to criticisms which undermine its credibility, and are here applied to both its positivist and its pragmatist varieties (Section 5). (However, readers anxious to proceed at once to belief in creation and its meaning, irrespective of the verificationist challenge, may prefer to move directly to Section 6, 'Creation and Analogy'.)

Building on the work of earlier empiricists such as David Hume and positivists such as Ernst Mach, the Vienna Circle of the 1920s and 1930s sought to fashion a comprehensive approach to philosophy in which acceptable ideas were confined to those based on sensory experience, and ideas not related to experience were deemed to be both metaphysical and nonsensical. The shared belief of this group was that science could encompass all proper questions within its scope: as Rudolph Carnap maintained, 'There is no question whose answer is in principle unattainable by science'.[6] They further believed that the role of philosophy is to propound principles of meaning which restrict meaning (beyond definitions and

---

5 See Hilary Putnam (2002), *The Collapse of the Fact/Value Dichotomy and Other Essays*, pp. 24–7; William Alston (2003), 'Religious Language and Verificationism', in Paul Copan and Paul K. Moser (eds) *The Rationality of Theism*, 17–34.

6 Rudolph Carnap [1928] (1967), *The Logical Structure of the World*, p. 290. Despite sympathy for such views, one member of the Vienna Circle, Moritz Schlick, held an illuminating stance on the meaning of life: see Chapter Nine, note 48 (below).

truths of mathematics and logic) to what is empirically verifiable, and consign metaphysics to meaninglessness.

Like their predecessors, these philosophers traced all human ideas to sensory impressions, plus necessary truths concerning the resulting notions. This approach gave rise to a conclusion of David Hume about works of metaphysics, a conclusion with which they were in broad sympathy: 'Does it contain any abstract reasoning concerning quantity or number? No. Does it contain any experimental reasoning concerning matter of fact and existence? No. Commit it then to the flames: for it can contain nothing but sophistry and illusion.'[7] The huge success of the scientific enterprise since the time of Hume reinforced their confidence that empirical methods could cope with all questions worth asking, and their impatience with entities that made no observable difference. Such entities, they held, were not only unknowable but incomprehensible. This was because of the nature of human understanding, and the limits of meaningful communication; for talk of such entities lies beyond these limits.

Thus according to one of their central claims, identified by Dummett as 'the first axiom of logical positivism': the meaning of a proposition is the method of its verification.[8] Accordingly, whether for a group of people or for a single investigator, the meaning of a claim consists in the ways in which it can be used in observing, experimenting or recognizing phenomena, and where there is no method of verification, an expression ceases to have a meaning. As A.J. Ayer expressed this:

> We say that a sentence is factually significant to any given person, if, and only if, he knows how to verify the proposition which it purports to express – that is, if he knows what observations would lead him, under certain conditions, to accept the proposition as being true, or reject it as being false.[9]

While objectors might be able to question such claims by finding Ayer's requirement unduly demanding or by specifying other kinds of use for expressions besides verification, the axiom mentioned by Dummett in any case conveys a key element in the motivation of the logical positivists, the belief that meaning and understanding depend on what language-users can do to verify units of language, and that what is unverifiable is therefore empty verbiage. As their pragmatist cousins might have expressed the matter, language that cannot be verified makes no practical difference, and may as well be abandoned. But unlike the pragmatists, the logical positivists regarded the kinds of experiences relevant to verification

---

7 David Hume (1975), *Enquiries Concerning Human Understanding and Concerning the Principles of Morals*, third edn, ed. L.A. Selby-Bigge, rev. P.H. Nidditch, *Human Understanding*, Sect. XII, Part III, p. 165.

8 Michael Dummett (1992), 'The Metaphysics of Verificationism', in L.E. Hahn (ed.), *The Philosophy of A.J. Ayer*, 129–48, p.130.

9 A.J. Ayer [1936] (1946), *Language, Truth and Logic* (2nd edn), p. 35.

as confined to those of the five senses, as our only means of accessing the world of shared experience.

The Principle of Verification now flows readily. Ayer's provisional expression of it runs like this: a sentence has literal meaning if and only if the proposition it expressed was either analytic or empirically verifiable.[10] The word 'proposition' here was replaced in the second edition of *Language, Truth and Logic* by 'statement', defined such that every grammatically significant sentence expresses a statement, whether literally meaningful or not; this move was supposed to avoid the problem that meaningless sentences fail to express propositions, capable of being either true or false, or therefore of being verified.[11] There may in fact be a deep problem here, since Ayer still has the related problem of specifying which 'statement' we are supposed to verify, and yet appears to need to require that the sentence can be known to be meaningful before this process can begin. But let us waive this problem for present purposes, and focus on the key terms 'analytic' and 'empirically verifiable'.

'Analytic' was used by Ayer and his fellow-positivists of propositions or statements that are necessarily true in virtue of the meanings of the terms employed in their expression, such as, for example, the claim that all sisters are female siblings. Subsequently the distinction between 'analytic' and its contradictory 'synthetic' was to be questioned by W.V.O. Quine in 'Two Dogmas of Empiricism',[12] but the questioning of a distinction is far from showing that there are no clear cases to which it applies, and for present purposes we may be content to accept Ayer's usage. The phrase 'empirically verifiable', however, raised problems for Ayer and his colleagues, depending on whether conclusive verification or some weaker form of verification was intended.

Conclusive verifiability seems the criterion best suited to upholding the verificationist axiom that the meaning lies in the method of verification. But it also turns out to exclude central features of scientific discourse, such as sentences involving dispositional terms (because they imply counterfactuals – claims such as 'If I were you, I would be elsewhere' – which are far from conclusively verifiable), and, worse still, expressions of laws of nature such as Darwin's principle of evolution by natural selection (because they imply not only counterfactuals but also predictions about the future, which are clearly unverifiable beforehand). Since the 'strong' version of the Verification Principle makes much of science meaningless (and much of history too), Ayer and others felt constrained to endorse a 'weak' version: a proposition 'is verifiable in the weak sense, if it is possible for verification to render it probable';[13] what we need to ask of a sentence is whether any observations would be relevant to the determination of its truth or

---

10 Ayer, ibid., p. 5.

11 Ayer, ibid., p. 8.

12 W.V.O. Quine (1953), 'Two Dogmas of Empiricism', in Quine, *From a Logical Point of View: Nine Logico-Philosophical Essays*, 20–46.

13 Ayer, op. cit., p. 9.

falsity.[14] But this weakening of the Principle not only introduces all the problems of induction and reasoning about probability, as advanced by Hume (problems that were to disincline philosophers such as Karl Popper from appealing to verification of any kind), but has also been argued to readmit both theoretical entities (some of them precluded by some of the positivists) and some of the claims of metaphysicians as well. Despite valiant attempts, Ayer never succeeded in expressing the weak version of the Verification Principle in a way that both served his basic purpose and avoided counterproductive problems such as these. The unsuccessful struggle of the logical positivists to overcome this difficulty is ably depicted in C.J. Misak's *Verificationism*, and was to be recognized by Ayer himself.[15]

But the difficulty of formulating the Verification Principle in a watertight manner is a flimsy and a vulnerable basis for resisting its implications, since the possibility of an improved draft of the Principle has always seemed capable of overcoming this problem at any moment. Hence the radical challenge of verificationism to metaphysics and theology remains worthy of attention. For religious believers of a theistic kind, in particular, characteristically make claims about a creator whose being and agency transcend empirical experience, and these are claims that fall well beyond the scope of science. Indeed these claims concern an agency which makes possible the phenomena that science studies. Even if some versions of the Verification Principle in the weak sense of 'verification' accidentally allow talk of such a being, the central thrust of the Principle is that such claims, provided they are not analytic, make no verifiable difference, and must therefore be rejected as meaningless. (It should be added at once that this challenge is not distinctively addressed to the theory of 'creationism', since it attacks equally the theism of those many theists who reject this stance, and accept Darwinism, but continue to believe in creation.) Ayer's 'Critique of Ethics and Theology' in *Language, Truth and Logic* shows how trenchant and far-reaching the challenge remained, even at a stage (as when the second edition was published in 1946) when some of the problems about strong and weak verification were already apparent. Indeed the same challenge has continued to be presented by later generations of philosophers in recent decades.[16]

Some interpretations of religious language, as Ayer remarked, are immune from this challenge, including those that represent it as concerning natural forces (the wind or the oceans or the sexual drive) or sectors of the natural world

---

[14] C.J. Misak, op. cit., p. 77.

[15] C.J. Misak, ibid., pp. 70–82; A.J. Ayer (1973), *The Central Questions of Philosophy*, pp. 26–27.

[16] Ayer, *Language, Truth and Logic*, 'Critique of Ethics and Theology' (pp. 102–20); see pp. 114–20 on belief in God. More recent expressions of the same challenge can be found in Kai Nielsen (1971), *Contemporary Critiques of Religion*, p. 57, and (1982), *An Introduction to the Philosophy of Religion*, pp. 41–2; and Michael Martin (1990), *Atheism: A Philosophical Justification*, pp. 40–78.

(continents or currents or heavenly bodies). Here language superficially concerning deities relates in fact to observable factors, and can be empirically verified as such. Thus much of the language of animism (and some of that of pantheism too, as concerning the natural world as a whole) survives the challenge.[17] Further interpretations treat religious language not as involving declarative statements at all, but as instead embodying normative prescriptions about ways of life (an approach favoured by R.B. Braithwaite).[18] If religion is ethics in disguise, this brings it within the verificationist critique of ethics; and while Ayer took the view that the claims of ethics are unverifiable, and thus do not embody propositions, he allowed them to have a different kind of meaning, in which speakers either express their own feelings, or seek to evoke relevant feelings in others. This emotivist and non-cognitivist account of ethical discourse is presented in the early part of his 'Critique of Ethics and Theology'.[19] (Ayer's finding room for prescriptions proves important for the status of his own theory, a theme to which we shall return.) Provided that religious believers abandon cognitive claims, verificationists can allow them emotive meaning along these lines. But believers have usually been reluctant to endorse this interpretation; while making ethical claims, they have widely wished to make them alongside cognitive claims of a verification-transcendent nature, such as claims concerning God and creation.

To these cognitive claims, the verificationist challenge remains. For to the extent that religious language purports to concern what transcends possible sensory experience, the Verification Principle would seem to construe it as literally meaningless.

Section 3: Some Implications of Verificationism for Religious Language

Thus the implications of the Verification Principle for the language of creation, and for theistic religions in general, are drastic. But does this consign religious believers to any different a plight from adherents of other widespread beliefs? And do the same implications attach to that older and more measured version of verificationism, pragmatism?

Strong verificationism, after all, (verificationism, that is, requiring language to be either analytic or conclusively verifiable if it is to be meaningful) implies that a great deal of scientific and historical language is meaningless too. In addition, it says the same about ethical and aesthetic language, with the proviso that such language may have emotive meaning; but then, the ethical and aesthetic aspects of religious language receive exactly the same treatment. The languages of theoretical entities, whether subatomic particles, the ether, essences, plastic

---

[17] Ayer, ibid., p. 116.

[18] See R.B. Braithwaite (1955), *An Empiricist's View of the Nature of Religious Belief*.

[19] Ayer, *Language, Truth and Logic*, pp. 102–14.

natures or Bergson's *élan vital*, are treated as meaningless alike; and the same goes for Freud's superego, ego and id, and Marx's dialectic of history. Indeed either very little (perhaps sentences like 'It seems to me as if I were seeing something red') satisfies this requirement, apart from analytic sentences, or maybe nothing at all (as would be the case if no claim is immune from questioning and thus conclusive). Thus the language of creation would simply be categorized with virtually every utterance of any interest. But the human community is unlikely to abandon interesting language.

But with weak verificationism, matters are different. Verification is no longer regarded as conclusive, but as involving (for example) language having observable implications, when conjoined with other sentences which do not have such implications alone; language of this kind would be held meaningful alongside analytic sentences. This example (borrowed from the first edition of *Language, Truth and Logic*) is in fact unsatisfactory, since any nonsense sentence (such as 'The Absolute is green') now becomes meaningful if conjoined with a conditional sentence of which the same sentence is the antecedent and an empirical fact is the consequent (like 'If the Absolute is green, there are ten people in this room'.) But we can imagine requirements to plug this loophole, and consequently the emergence of a revised Verification Principle that would recognize language the implications of which could make an observable difference (language about medium-sized observable objects like tables and language about electrons, for example), but not language about the ether, the id, or the dialectic of history.[20] Ethical and aesthetic language would be treated as before (as literally meaningless but having emotive meaning), and language about deities which was not itself ethical or aesthetic would be treated as meaningless without qualification. Historical language, however, would pass muster, together with historical elements of religious language of an empirical kind (e.g. 'He was crucified under Pontius Pilate'). But apparently historical claims concerning transcendent entities (e.g. 'He was conceived of the Holy Ghost') would not.

Language about creation, together with most religious language, would thus be clustered with language that cannot be claimed to fall within the disciplines of science or of history. But being neither scientific nor historical is not normally considered sufficient to make a remark literally meaningless, any more than ethical language is so regarded, or the language of psycho-analysis. Weak verificationism seems, despite itself, to have managed (or rather, to have nearly managed, in view of no adequate version of the Verification Principle ever having been supplied) to supply a possible criterion for language to be scientific, or perhaps to be either scientific or historical. But many purposeful utterances are never held to fall under either of these disciplines, and yet to be meaningful within their own sphere. And

---

20 William P. Alston, however, has recently argued that no way to plug this loophole is likely to be found. See Alston, 'Religious Language and Verificationism' (2003), in Paul Copan and Paul K. Moser (eds) *The Rationality of Theism*, 17-34, pp. 28–30.

what this suggests is that we should consider whether the Verification Principle should be accepted at all, or radically revised (see the coming section).

However, another form of verificationism adopted a different stance on these matters. For pragmatists have adopted much more liberal interpretations of the range of experience, the nature of verification and the possibilities of metaphysics. C.S. Peirce, for example, enlarges the scope of experiential consequences relevant to verification so as to include expectations in theory, ones that might emerge when we are studying diagrams rather than receiving sensory inputs or making physical measurements. As Misak remarks, this overcomes the problem for adherents of the Verification Principle that lines of $\sqrt{2}$ units of length cannot be physically measured (a problem arising wherever irrational numbers are at issue). As long as our experience when studying diagrams could surprise us (as is the case with the hypotenuse of a right-angled triangle), it is to be counted as potential verification.[21] This recognition of theoretical expectations demonstrable on diagrams considerably broadens the scope of discourse to be recognized as meaningful, admitting (for example) much of the work of economists, geographers and other social scientists, as well as that of geometers. Effectively it allows most academic disciplines to advance theories and test them against expectations of what they imply.

Peirce was even prepared to extend this recognition to some metaphysics, which he regarded as in principle an observational science. The phenomena of metaphysics are less obvious than those of science, being 'harder to see, simply because they surround us on every hand; we are immersed in them and have no background against which to view them'.[22] Yet metaphysical hypotheses must still have observational consequences. 'Nothing comes from nothing' could well serve as a relevant example, having been held long ago by Lucretius to have numerous such implications.[23] However, according to Peirce the backward condition of the discipline of metaphysics is to be blamed on the fact that 'its leading professors have been theologians'.[24] Thus at least some theological beliefs were held to lack observational consequences (including Leibniz's theory of pre-established harmony),[25] just like Bertrand Russell's theory that the world might have been brought into being five minutes ago, complete with fossils and geological strata, a theory which generates exactly the same empirical expectations as standard theories about the past, and is to be rejected (Peirce held) because it makes no practical difference and coheres less well with the rest of our beliefs.[26] (But, as

---

[21] Misak, op. cit., p. 111.

[22] C.S. Peirce (1935), *Collected Papers of Charles Sanders Peirce*, (8 vols; vols 1–6) ed. C. Hartshorne and P. Weiss, vol. 6, p. 562; Misak, ibid., p. 119.

[23] Lucretius (1947), *De Rerum Natura*, ed. and trans. Cyril Bailey (3 vols), vol. I, p. 205: 'nothing can be brought to being out of nothing'.

[24] Peirce, op. cit., vol. 6, p. 3; Misak, op. cit., p. 118.

[25] Misak, ibid., p. 99.

[26] Misak, ibid., p. 123.

a critic might retort, this hardly makes it meaningless, or even, as Peirce would more probably have held, as blocking the road to inquiry.[27])

With relation to the standard hypothesis about the past (the one opposed to Russell's) and comparable theories, Peirce upheld a third kind of reasoning beside deduction and induction: that of abduction, or reasoning to the best explanation. If a hypothesis best explains some surprising phenomenon, then we have reason to admit it into our enquiries, and to apply deductive and inductive methods to test it and its coherence with other theories.[28] While there are large areas of metaphysics which this does not legitimate (talk of the Absolute probably comprising a central instance), it does seem to legitimate discussion of metaphysical explanations such as those purporting to be the best explanation of all sensory experience and of its possibility, such as belief in divine creation. Peirce, who himself produced an argument for the existence of God (see Section 5 below), could well have agreed.

Section 4: Ayer's Subsequent Stances

Without accepting the objections of critics to verificationism, Ayer became increasingly forthright in recognizing that the Verification Principle had never been adequately expressed, and that all attempts to express it successfully had been abortive. This may explain his not appealing to it when discussing 'The Claims of Theology' in his Gifford Lectures of 1972 and in their published version, *The Central Questions of Philosophy* (1973). Instead, he confined himself to expressing the possibility that talk about God is incoherent: 'There may be some doubt whether the predicates that are ascribed to this one God are all of them meaningful or mutually consistent.'[29] But in support of this carefully articulated scepticism he supplied particular grounds, such as the possibility that if God is a disembodied person, he must be located in time and not be timeless. After ventilating several such problems, Ayer proceeded critically to analyse the various arguments for the existence of God, prefacing his able critique of the ontological argument with the recognition that 'The idea that God necessarily exists is worth pursuing ...'.[30] Thus the reader who remembers his avowal in an earlier chapter of the inadequacy of all attempts to express the Verification Principle[31] could readily interpret him as holding that language about God is incoherent only on occasions when special factors make it so, and that this makes the case for atheism (itself now treated as a coherent position, as it was not in *Language, Truth and Logic*) worth arguing.

---

27 Misak, ibid., p. 99.
28 Misak, ibid., p. 123.
29 Ayer, *Central Questions of Philosophy*, p. 212.
30 Ayer, ibid., p. 213.
31 Ayer, ibid., p. 27.

The way in which Ayer got to grips with the theistic arguments suggests that by now he regarded much of their language as meaningful, while rejecting their conclusions. It is virtually impossible to interpret his text as written by someone just playing along with the coherence of such language for the sake of argument, for incoherent sentences fail to express propositions and cannot be regarded as conveying arguments, whether valid or invalid. Yet Ayer remained a verificationist, as he affirmed much later in his reply to Michael Dummett, where he endorsed Dummett's suggested encapsulation of verificationism in the words 'The meaning of a statement is the method of its verification'.[32] So Ayer's eventual position is best construed as adherence to a weak version of verificationism, combined with acceptance that this does not preclude all language about God, or indeed all metaphysical discourse either. In the latter connection, he was prepared to recognize meaning in the kind of metaphysics which embodies semantic theories, as in Plato, Leibniz and Spinoza, but not the utterances of Heidegger and Derrida, which he continued to regard as nonsensical.[33]

This interpretation proves of assistance in the matter of understanding Ayer's accounts of his near-to-death experience of 1988, when for four minutes his heart stopped beating. In his first written account, published in the *Sunday Telegraph* (28 August 1988), he wrote as follows:

> The only memory that I have of an experience, closely encompassing my death, is very vivid. I was confronted by a red light, exceedingly bright, and also very painful even when I turned away from it. I was aware that this light was responsible for the government of the universe. Among its ministers were two creatures who had been put in charge of space. These ministers periodically inspected space and had recently carried out an inspection. They had, however, failed to do their job properly, with the result that space, like a badly fitting jigsaw puzzle, was slightly out of joint.

In his dream-experience, Ayer had tried to alert the ministers of space, and, failing that, the ministers of time, so as to put matters right, but was unable to attract their attention, despite desperate attempts that lasted until the experience ended.[34]

In a scholarly yet brilliant interpretation of Ayer's dream, which takes its cue from his spoken remark immediately after this episode ('My thoughts became persons'), Abigail L. Rosenthal puts forward the theory that within this experience Ayer found himself obliged to live with the implications of his own theory of space, a theory that he had long feared to be defective.[35] Fortunately there is no need to discuss the details here of Ayer's theory of space or of Rosenthal's

---

[32] Ayer, 'Reply to Michael Dummett', in Hahn, p. 152.

[33] Ayer, ibid., p. 150.

[34] Ayer (1988), 'What I Saw When I Was Dead', *Sunday Telegraph*, 28 August; reprinted as 'Still More of My Life' in Hahn, 44–50. The quoted passage is from p. 46.

[35] Abigail L. Rosenthal (2004), 'What Ayer Saw When He Was Dead', *Philosophy*, **79**, 507–31.

incisive commentary. What is of current relevance is the use of religious language in the words, 'I was aware that this light was responsible for the government of the universe'.

This report can of course be interpreted as short for 'In my dream I believed that this light was responsible for the government of the universe'; and it would be a fallacy to infer that if someone is reported as believing something to be coherent, then it is coherent, or that the person reporting this accepts its coherence, even if the person with the belief and the person reporting are identical. Yet Ayer (in his dream) took the government of the universe just as seriously as the disjointedness of space, and after recovering went on to make the comment: 'It is conceivable that one's experiences in the next world, if there are any, will supply evidence of a god's existence, but we have no right to presume on such evidence, when we have not had the relevant experiences.' (This reads like a qualified recognition of John Hick's advocacy of the possibility of post-mortem or 'eschatological' verification.[36]) He added that it is possible to believe in post-mortem experiences while being an atheist, as in the cases of J.E. McTaggart and C.D. Broad.[37] He also discusses whether his experience had been 'delusive' or 'veridical';[38] and this suggests that he considered, when writing this, that it is in principle possible that someone or something is indeed 'responsible for the government of the universe', although he also held that there is inadequate reason to believe this to be the case.[39] Ayer accepted that the parallel near-to-death experience of a friend's mother supplied some slight evidence that his own experience might have been veridical.[40] Its being 'veridical' would imply that there really is someone or something responsible for the government of the universe, and would also be compatible with the universe being created. In a further article, Ayer discussed the relative probability of bodily resurrection (improbable from his perspective) and of reincarnation (slightly less improbable), either of which could in theory sustain experiences parallel to those of his dream.[41]

Verificationism, then, does not preclude the coherence of belief in the government of the universe, or in its being created. While early verificationism had seemed to preclude these beliefs, its apparent inexpressibility in a credible version that would have consigned them to meaninglessness leaves them intact as metaphysical possibilities, as far as verificationism is concerned. As we have seen from Ayer's own experience, they may also be epistemological possibilities, although he did not consider them cogent as such. (Hick gives reasons why a

---

[36] John Hick (1971), 'Theology and Verification', in Basil Mitchell (ed.), *Philosophy of Religion*, 53–71.

[37] Ayer, 'Still More of My Life', p. 48.

[38] Ayer, ibid., p. 47.

[39] Ayer, ibid., p. 48.

[40] Ayer, ibid., p. 47.

[41] Ayer (1988), 'Postscript to a Postmortem', *Spectator*, 15 October; reprinted as part of 'Still More of My Life' in Hahn, pp. 50–53.

theistic believer who has relevant additional experiences might regard them otherwise.[42]) Hence even if the objections to verificationism to be discussed in the coming section are misdirected, verificationism does not of itself supply grounds for regarding belief in creation as meaningless.

Section 5: Some Objections to Verificationism

We turn now to two problems for verificationism, both touched on in the second section above. One raises the issue of how to specify what we are supposed to verify, when the meaningfulness of the relevant sentence is under consideration and cannot yet be taken for granted, or therefore its meaning known. The other concerns the status of the Verification Principle itself, mentioned previously in the context of prescriptions.

The clearest expression of the first objection was penned by J.L. Evans in 'On Meaning and Verification':

> But it is of the greatest importance to realise that [a sentence's] unverifiability is the consequence and not the cause of its being meaningless. It is not the case that we first attempt to verify or falsify a sentence and then, when we fail, conclude that the sentence is meaningless. We realise that a sentence is meaningless and in doing so realise also that it is impossible to verify or falsify it by observation.[43]

As Evans concludes, this consideration is sufficient to overthrow the Verification Principle as supplying a criterion of meaningfulness or of meaning. If asked why we cannot first attempt verification and later proceed to form a conclusion about a sentence's meaningfulness, he could explain that in order to consider a statement or proposition for verification we need to understand it, and to understand it we already need to grasp what it means. Processes of verification thus presuppose the meaningfulness of the relevant sentences, and their understanding on the part of would-be verifiers. Hence verifiability cannot conceivably be a criterion of the meaningfulness of the relevant sentences. Indeed there would be no way of describing those statements or propositions which we would initially understand and regard as meaningful sufficiently to attempt to verify them but subsequently conclude never to have been meaningful in the first place. Self-contradiction thus attends any attempt to describe verifiability as a criterion of meaning.

What can be verified, it should be added, must be capable of being true or false, and will therefore be a proposition (whether or not it is asserted, and thus comprises a statement). But many sentences are capable of expressing an unlimited range of propositions. Thus the sentence 'She hit him with it' could refer to

---

[42] Hick, op. cit., pp. 70–71.

[43] J.L. Evans (1953), 'On Meaning and Verification', *Mind*, **62**, 1–19, p. 12. This article is praised as 'excellent' and 'unusually perceptive' by Karl Popper at Popper (1963), *Conjectures and Refutations*, p. 41, n. 7.

Margaret Thatcher hitting a colleague with a handbag, to Boudicca smiting a Roman with the wheel of a sword-chariot, or to a submarine striking a swimmer with its conning-tower. Our identifying which proposition is intended depends on our grasp of the sentence, of its occasion of utterance, and on the intended likely context; only thus can we discover the references of 'She', 'him' and 'it'. (Quinean problems relating to the identification of the meaning of sentences are discussed in Chapter Three, Section 1 below, and can be set aside for the present.) To suggest, as Ayer does,[44] that there is some single 'statement' which corresponds to a sentence whether it is meaningful or meaningless, and that this is the unit that may or may not be verified and that may thus be found to have or to lack meaning, is to confuse several issues (not least that of what is involved in making a statement: see Chapter Three below); for present purposes, it confuses the meaning of a sentence (which need not be true or false at all, if not used in communication about a specific context) with each and every proposition of the infinite set that could be expressed by that sentence. If we can get as far as verifying one of the propositions that we interpret the sentence as expressing, then we already presuppose the meaningfulness of that sentence.

A version of the same objection can also be deployed if a sophisticated verificationist attempts to cope with the problems mentioned in the last paragraph by proposing that an indicative sentence is meaningful if and only if it can be used to make a verifiable statement.[45] Once again, as a precondition of our being in a position to know whether a statement is verifiable, we need already to know both the meaning of the expression in which it is couched, the occasion of utterance and the likely intended context; otherwise the reference(s) it makes will be unclear, and any statement(s) that may be made will not be identifiable. What this serves to bring out is the fact that the meaning of the relevant expression or sentence must be discoverable before the issue of verifiability can be addressed, and also the further fact that in the event of there being nothing to which its referring expressions succeed in referring, there may still be no statement made (whether verifiable or not) at all. (Indeed far from all meaningful indicative sentences are usable to express verifiable statements. Some, for example, could only concern people of the distant, irretrievable past, about whom nothing can any longer be verified. Anti-realists dispute whether such sentences can be true, as will emerge in Chapter Two below, but not whether they are meaningful.) Exactly the same can be said if the proposal claims, in the light of discussions of falsifiability (such as that of Chapter Three below) that an indicative sentence is meaningful if and only if it can be used to make a statement that is either verifiable or falsifiable. Hence even this sophisticated revision of verificationism proves to be both misguided and self-undermining.

---

44 Ayer, *Language, Truth and Logic*, p. 8.

45 Alston (an opponent of verificationism) has recently offered such a revision, in 'Religious Language and Verificationism' at pp. 19–20, but that version is open to the objection expressed in the current paragraph, as well as to those that Alston endorses.

Thus the objection presented by Evans undermines the Verification Principle, which assumes that sentences can be meaningful because either they or what they express are verifiable. To the extent that pragmatists such as Peirce make verifiability a criterion of meaning (even as one among others) they too are open to this criticism. The verifiability of a proposition may confirm the meaningfulness of a sentence that expresses it, but since it presupposes that the sentence has a meaning, it cannot form part of a necessary or sufficient condition for meaningfulness or even one disjunct of such a condition. The Verification Principle and comparable versions of verificationism turn out fundamentally to confuse semantic issues of the meaning of expressions with epistemological ones of verifiability. While sentences are either meaningful or meaningless, it is propositions (some of which are asserted and are thus statements) but not sentences which are, as such, eligible in some circumstances for verification. (Nor will it do to suggest in reply that this distinction is not viable since we lack criteria for identifying sameness of meaning, and hence to identify propositions. If this were true, the everyday practices of translation and interpretation would be impossible, and almost certainly that of conversation too.) The axiom of verificationism that the meaning of a sentence is (or depends on) the method of verification is likewise a hopeless confusion. It follows that no version of verificationism shows metaphysical or theological sentences to be meaningless, where verificationism is understood as making verifiability either a condition or part of a condition of sentential meaning or of a sentence's meaningfulness. (The adjacent issue of the relation of the meaning of sentences to the truth-conditions of the propositions that these sentences can be used to express is discussed in Chapter Two, Section 3 below.)

The second main objection to the Verification Principle concerns the status of that Principle itself. For according to that Principle it must, if meaningful, be analytic or else the proposition it expresses must be verifiable. But it does not express an empirical proposition, and is thus ineligible for verification, and (as Ayer recognized) it is not an analysis of ordinary language either, as it does not reflect the ordinary sense of terms like 'meaningful'.[46] So if it is analytic, it amounts to no more than a new proposal or stipulative definition for the use of terms such as 'meaningful'; and if not, it is meaningless on its own terms. Objections of this kind have been raised ever since *Language, Truth and Logic* was first published. But they have also been resisted, and not only by Ayer's followers.

J.L. Evans, for example, replies that 'we should not expect the Principle of Verification itself to be subject to the criteria which it lays down governing meaningful statements. We do not expect a weighing-machine to weigh itself.'[47] We certainly do not; but a principle could still fall within its own scope, and this Principle claims to apply to sentences that express propositions or statements, and at the same time is itself expressed in a sentence apparently of this kind.

---

46 A.J. Ayer (1992), 'Reply to Michael Dummett', in Hahn, 149–56, p. 149.

47 Evans, op. cit., p. 3.

Michael Dummett later argued that the Principle has to be construed as limited in scope; it describes one fragment of our language which does not contain the theoretical terms of the description, and is expressed in another that includes those terms; in this way the contradictions of self-reference can be avoided. Dummett allows that the words 'meaning' and 'verification' belong to ordinary English usage, but suggests that within the theory they figure 'solely as theoretical terms'.[48] While it might be objected that we can only understand 'meaning' and 'verification' as used within the theory if they carry similar meanings to those they have in ordinary discourse, Ayer seems in any case to have endorsed Dummett's approach, thus retracting the claim that the Verification Principle applies to indicative sentences in general.[49] At the same time Ayer recognized that this blunts the cutting edge of the Principle. For the Principle must now be construed as a prescription, and this raises the legitimate question, 'why should the principle be obeyed?' Ayer recognizes that in *Language, Truth and Logic* he 'evaded this awkward question by defying my critics to come up with anything better'. To this he adds an echo of Dummett's further suggestion that such prescriptive definitions may teach us something, albeit something not about the world but about concepts.[50] For example, as Herbert Feigl claimed in 1963, if the Principle is construed as a norm about meaning, we can regard it as precluding unanswerable questions,[51] questions unanswerable presumably because any answer would be meaningless.

But this reason begs the question of whether the precluded questions really are unanswerable and their answers meaningless. Maybe some are, but the positivists are in no position to appeal to the Verification Principle in support of this view when seeking reasons for adopting that Principle as a norm or prescription. Thus in the absence of other reasons, Feigl's argument fails. By now the second objection may be modified to the objection that the Verification Principle, construed now as a proposed definition of meaningfulness, is nothing better than an arbitrary prescription, lacking reasons for its adoption; and this objection, I suggest, holds water. However, it could be regarded as partially dependent on the previous objection, which embodies strong grounds for rejecting the Principle, and thus strengthens the case against accepting it as a prescription.

---

48 Michael Dummett, 'The Metaphysics of Verificationism', in Hahn, 129–48, pp. 129–33.

49 Ayer, 'Reply to Michael Dummett', p. 149.

50 Ayer, ibid., pp. 149–50. More recently, Alston has rejected as implausible the interpretation of verificationism as prescriptive, and retained the standard interpretation, apparently disregarding the seriousness of the charge (rehearsed in the same paragraph) that on this interpretation verificationism fails its own requirement and undermines itself. See 'Religious Language and Verificationism', at pp. 21–2.

51 Herbert Feigl (1963), 'Physicalism, Unity of Science and the Foundations of Psychology', in P.A. Schilpp (ed.), *The Philosophy of Rudolph Carnap*, 227–67, pp. 237–8.

To what extent is the verificationism of Peirce subject to the same criticisms? With regard to the second objection, Peirce took the view that pragmatist claims about meaning are themselves subject to pragmatist criteria of meaningfulness,[52] and would presumably not welcome a Dummett-style division of language into the fragment to which they apply and that to which they do not, not even to avoid paradoxes of self-reference. Since the claims of pragmatist theory themselves generate theoretical expectations, just as pragmatists expect meaningful first-order language to do, they could be held to survive this objection.

But since these claims (like those of Ayer and the positivists) must in any case be regarded as having a prescriptive role, the real issue here is rather that of whether there is good reason to endorse these claims. And these claims could reasonably be held to exhibit at least a tension, if not a contradiction, when they simultaneously recognize reasoning backwards to the best explanation (abduction) and reject as meaningless the theories of theologians in general.[53] For there are arguments from the fine-tuning of cosmic variables to the existence of life which conclude that the existence of an omnipotent and omniscient creator supplies the best explanation of these phenomena. These arguments will receive more detailed consideration later (in Chapter Five, Sections 3 and 4), but any suggestion that their conclusion is meaningless would serve to block inquiry, and should therefore be rejected on pragmatist grounds. Yet Peirce's semantic theories could be interpreted as precluding as meaningless just such conclusions (as they did in the case of the pre-established harmony of Leibniz, misguided as this principle is usually taken to have been). This suggests that Peirce's theories should either be rejected or radically revised. However, Peirce was in fact himself a theist, argued against positivism, and supplied a 'pragmaticist' argument for belief in the reality of God.[54]

Nevertheless, as mentioned above, Peirce's pragmatism is vulnerable to the first objection, since he makes verification one of the criteria of meaningfulness. As was shown above, this approach involves serious confusion. Pragmatists could perhaps reply by emphasizing other key tenets, such as their views on truth or on reality. Thus semantic anti-realism (to be discussed in the next chapter) might comprise their next line of defence. But to fall back on this defence would involve abandoning the verificationist theory of meaningfulness under discussion here; and this remains a theory which would seem to collapse in face of our first objection.

---

52 Misak, op. cit., p. 109.

53 Peirce, op. cit., vol. 6, p. 3; Misak, op. cit., p. 118.

54 Charles S. Peirce (1958), 'Notes on Positivism', in Philip P. Wiener (ed.), *Charles S. Peirce: Selected Writings*, 137–41; and Peirce (1908), 'A Neglected Argument for the Reality of God', *The Hibbert Journal*, **7**, 90–112, reprinted in Peirce Edition Project (ed.) (1998), *The Essential Peirce: Selected Philosophical Writings*, vol. 2, 434–50.

Section 6: Creation and Analogy

Verificationism, then, seems not to pose problems for the meaning of religious language. But there are related problems for making sense of such language. How can we understand talk of what lies beyond experience when our language is perforce derived from everyday experience, concepts and purposes? And what does it mean to talk of creation in the first place? These are questions about meaning, and thus precede questions of justification and truth. Rather than concerning whether God exists, they concern whether talk of God makes sense. A preliminary treatment of these questions (to be developed further in later chapters) is now in place.

In a dialogue entitled 'Creation', Donald MacKinnon (a Christian) and Antony Flew (then an atheist) agreed that belief in creation is centrally belief that the entire material universe, humanity included, is dependent on God.[55] While this dependence is compatible with the world having had a beginning, the belief that its past is finite is not essential to belief in God as 'the author of the world', as opposed to belief in its dependence on God at all times.[56] But an implication of this dependence is that it is dependent on nothing else, and was not, for example, made out of anything existing already or anything existing independently of God.[57] (This accounts for the traditional belief of theists that creation is *creatio ex nihilo*, or 'a bringing into being out of nothing'). Implicitly, to be God is to have the power, knowledge and wisdom to bring into being anything that can (without contradiction) be brought into being; and, according to theistic belief, it is (as was mentioned above) God's agency which makes possible the phenomena that science studies. (These themes are discussed in greater detail in Chapter Four below.) Flew and MacKinnon further agreed that belief in universal dependence on a single divine author eventually fostered and supplied some kind of rational support for at least the beginnings of scientific inquiry and the empirical search for universal regularities pervading creation, a thesis previously argued by Michael Foster.[58]

Recently, Paul Copan and William Lane Craig have contended that belief in creation is to be interpreted as essentially involving belief in the universe having been generated by God a finite time ago.[59] Here, however, I follow the contrary

---

55 The phrase 'material universe' is used here and elsewhere interchangeably with 'physical universe'; there is no suggestion that matter is more fundamental than energy, or that energy is any less dependent on God than matter is.

56 Antony Flew and D.M. MacKinnon (1955), 'Creation', in A. Flew and A. McIntyre (eds), *New Essays in Philosophical Theology*, 170–86, pp. 172–4. See also Richard Taylor (1974), *Metaphysics* (2nd edn), p. 107.

57 Flew and MacKinnon, p. 172; 2 Maccabees 7:28 suggests that Judaism understood creation in this way before Christianity did.

58 Flew and MacKinnon, pp. 176–8; Michael Foster (1934), 'The Christian Doctrine of Creation and the Rise of Modern Natural Science', *Mind*, **43**, 446–68.

59 Paul Copan and William Lane Craig (2004), *Creation out of Nothing: A Biblical, Philosophical and Scientific Exploration*.

stance of Ian Barbour, who argues, along the lines of MacKinnon and Flew, that the central theological affirmations of Genesis and other biblical passages about creation concerned the dependence of the world on God, the world's orderliness and intelligibility, and God's sovereign freedom, and not a claim about cosmic beginnings capable of either tallying or clashing with scientific findings.[60]

However, talk of God as author or as agent, just like language about the government of the universe (introduced above in Ayer's description of his dream), cannot be taken in the ordinary sense of those terms, and the same applies to talk of God's will or purposes. For we cannot ascribe these terms to God on the same basis as that on which they are applied to human authors or governments or designers. Whether this comprises a complete obstacle to understanding such talk about what lies beyond our experience will further be discussed in the coming chapter. But it is relevant here to introduce one traditional account of how it is possible to apply to God predicates normally applied to people, an account presented in the same dialogue by MacKinnon.

MacKinnon introduces two accounts of the analogical use of language, intended to show how language about God can avoid the twin pitfalls of being used in the same sense as it is used of humans (anthropomorphism) and being used in an altogether different but unspecified sense (equivocation). The Analogy of Attribution suggests that God is good in the sense of being the ultimate source or cause of goodness, and compares this (following Aristotle) with the way in which fresh air or seaside resorts can be said to be healthy as being causes of health. In the Analogy of Proportionality, however, God's goodness and other attributes are held to be related to God's nature in the same manner or ratio as human goodness is to human nature. With this second Analogy, the problem concerns whether we can know what God's nature is, a problem that has led some writers to abandon this entire approach.[61] This is probably why MacKinnon maintains that the Analogy of Attribution is fundamental, and 'gives you what you must take over into' the other Analogy.[62] While in itself, Attribution fails to capture the meaning of 'God is good' (for much more is usually meant by this than that God is the cause of goodness), it supplies us with what we need to operate the other Analogy, that is, God's nature as creator[63] (or, to avoid asserting too much in a theory of meaning, as potentially creator). God's goodness would be goodness at this level, and is related to God's other properties as human goodness is to

---

[60] Ian G. Barbour (1998), *Religion and Science: Historical and Contemporary Issues*, pp. 199–204. The relation of creation to time is further discussed in Chapter Four, Section 4 (below).

[61] Thus Frederick Ferré (1961), *Language, Logic and God*, Chapter 6, 'The Logic of Analogy'.

[62] MacKinnon in Flew and MacKinnon, op. cit., p. 182.

[63] Much light is thrown on the logic of creation and on what differentiates it from ordinary causation in Peter Geach (1969), 'Causality and Creation', in Geach, *God and the Soul*, 75–85.

characteristic human properties. This seems both a reasonable interpretation of MacKinnon's view, and also a tolerable theory of Analogy, relevant to divine authorship, agency, and governance as well as to God's goodness. It also involves that fragment of religious language which sets the level at which the rest applies being used literally; this granted, the rest can be interpreted on an analogical basis in accordance with the theory. And this seems a reasonable requirement, despite recent doubts and reservations.[64]

Considered in this way, the Analogy of Proportionality could be applied to many other contexts, where predicates of humans are being applied to animals, machines or systems. Ascriptions, for example, of intelligence to dogs or horses need to be understood as related to the nature of dogs or horses as human intelligence is to human nature. Similarly, ascriptions of intelligence to computers need to be understood in a similarly qualified manner (or not at all), and the same applies to corresponding ascriptions to global systems, as when intelligence is ascribed to Gaia (the system of planetary systems). In all these cases, we can, with some difficulty, make sense of the discourse on offer. This discourse in each case precludes certain implications that would be standard in the human case, but frequently illuminates the case under discussion nonetheless. Similarly the language of analogy can be applied to God in a manner which, I suggest, avoids both anthropomorphism and equivocation,[65] and allows issues such as whether a good God would permit the world's evils to be coherently discussed (as in Chapters Six and Seven below). To this language, objections concerning a supposed lack of grounds for belief in God's existence, raised in the dialogue by Flew, are beside the point, since the theory to which he was objecting is a theory of meaning.[66] (Fifty years on, now that he has concluded that some kind of deity probably does exist, albeit not the God of theism, he might well retract that objection.[67])

It is worth pausing to draw an implication of such analogical extensions of language – that the bounds of meaning and of understanding stretch much wider than many philosophical theories (including verificationism) permit. While meaning certainly has its limits, it is unclear that the deaf cannot understand

---

64 David Burrell (1973), *Analogy and Philosophical Language*, pp. 132–33; William Alston, 'Religious Language and Verificationism', 17–34, pp. 32–34.

65 See further Robin Attfield (1980), 'Religious Symbols and the Voyage of Analogy', *International Journal of the Philosophy of Religion*, **11**, 225–38, J.P. Ross (1981), *Portraying Analogy*, pp. 93–108, and Robin Attfield (1996), 'Too High a Theme? Of Finitude, Predication and Analogy', *Scottish Journal of Religious Studies*, **17**, 5–19. For another view, see Richard Swinburne (1992), *Revelation: From Metaphor to Analogy*, pp. 150–57.

66 Flew in Flew and MacKinnon, p. 182.

67 Antony Flew (2004), 'On Darwinism and Theology', *Philosophy Now*, **47**, August/September, 22; and Flew (2005), 'My 'Conversion'', *Think: Philosophy for Everyone*, **11**, Autumn, 75–84. See also Richard Carrier (2005), 'Antony Flew Considers God ... Sort Of', <http://www.secweb.org/asset.asp?AssetID=369>, most recently updated, January 2005, when visited in October 2005.

talk of pitch and timbre, or blind people talk of colours. Are blind students of literature, for example, unable to grasp Shakespeare's meaning when he writes:

> No, this my hand will rather
> The multitudinous seas incarnadine,
> Making the green one red?[68]

And beyond the sphere of what is clearly understood there is also the sphere of fragmentary understanding, where we infer meanings that are no more than partially determined by the half-seen or half-heard clues available. (I am not suggesting that fragmentary understanding is characteristic of religion more than other areas of human activity, but that it is widespread in them all.) Theories that understanding invariably involves the ability to verify or even to express this or that required item defy much common experience. Rather, waves of meaning from the seas of language[69] engulf us, lapping at our doors and windows, heedless of Canute-like requirements and prohibitions.

Yet Flew interestingly adds another facet of the theistic belief in creation, worthy of mention because of its relevance to the later chapters of this book. He writes concerning the insistence that this is God's world that it surely suggests 'that we should behave, as it were, as guests and borrowers: not as owners who have a right to do what they like with their own'. This is, he suggests, a moral implication for our conduct and attitudes,[70] (or at least it could be held to comprise, as we might re-express Flew's remark in Gricean terms, an implicature or contextual implication, if not an implication in the strict sense): a view which could be held to cohere with attitudes of stewardship, which are often associated with Old Testament passages such as 'The Earth is the Lord's' (Psalm 24:1). For the present, however, I intend to treat such attitudes of 'creatureliness' as a separate topic, and not to include them as integral to belief in creation, since believers in creation do not always adopt them, even though, historically, they have often formed crucial aspects of what gives life its meaning for theists and theistic cultures, and even though some philosophers, as will shortly be seen, treat these links as more than contingent ones. This is not to suggest, however, that belief in creation lacks a practical import. Indeed it is here that pragmatists might well find much of its meaning.

The current discussion of creation and analogical language could be taken to suggest that theistic language overcomes the challenge of verificationism, and becomes (or remains) a serious candidate for our acceptance. But this would be to reckon without recent anti-realist theories and stances that have evolved out of

---

68 William Shakespeare, *Macbeth*, II, ii, 60–62.

69 This phrase is derived from the title of Michael Dummett (1993), *The Seas of Language*.

70 Flew, in Flew and MacKinnon, p. 173; H.P. Grice (1989), 'Logic and Conversation', in H.P. Grice, *Studies in the Way of Words*, 22–40.

verificationism. For some of these theories further question the ability of human beings to understand claims that transcend possible experience. To these theories and stances I turn in the next chapter.

Chapter Two

Realism, Anti-realism and Religion

Section 1: Realism and the Roots of Anti-realism

In a work entitled *Religion and Philosophy*, Martin Warner introduces realism as follows:

> In general terms, realism is the doctrine that that which we encounter exists quite independently of us – although not necessarily in the form that we conceive of it – and that reality (whether physical, temporal, mathematical, mental, moral, religious or whatever) is therefore independent of our conceptions of it.[1]

Anti-realism, by contrast, involves a denial of this for any one (or more) of the specified fields. Realism and anti-realism, then, are metaphysical theories, and have most often been taken to concern matters of what there is.

However, as was mentioned in the previous chapter, Michael Dummett has argued that theories of meaning are fundamental in philosophy, and that the issue between realism and anti-realism turns on corresponding semantic theories, or theories of meaning and understanding.[2] Whether or not we grant that theories of meaning are fundamental, the importance of theories of meaning is pivotal for at least some of the fields just mentioned. As these include the philosophy of religion, it is crucial to consider the relevant theories of meaning here.

The relevance of theories of meaning can be illustrated by reference to a different field, that of moral discourse. Here realism, which maintains that right and wrong are objective properties, ones that, according to ethical cognitivism, can sometimes be known, stands opposed to non-realist theories, many of which reject belief in objective properties and moral knowledge (non-cognitivist theories), and construe the meaning of ethical terms as either projections of emotions or attitudes or as prescriptions. (As we have seen, A.J. Ayer adopted one of these theories of ethics.[3]) A different form of anti-realism is the error theory of J.L. Mackie, according to which the objective and categorical status of

---

1 Martin Warner (1992), 'Introduction' in Martin Warner (ed.) *Religion and Philosophy*, 1–21, p. 4.

2 Michael Dummett (1973), *Frege: Philosophy of Language*, p. 669. See also Dummett (1978), *Truth and Other Enigmas*, 145–65.

3 A.J. Ayer [1936] (1946), *Language, Truth and Logic*, 'Critique of Ethics and Theology' (pp. 102–20); see pp. 102–14 on ethics.

moral properties such as 'right' is such that all ascriptions of them are erroneous.[4] In this latter case, anti-realism has both a meaning-component (the meaning of 'right' etc.) and an ontological component (the claim that such properties apply to nothing in reality). But in the case of non-cognitivist theories such as emotivism and prescriptivism, anti-realism turns entirely on the relevant theory of meaning; and it is because of this, supposedly, that objective rightness is not to be found or cannot be known. In the field of ethics, then, anti-realism is not typically an ontological theory, nor therefore realism either, and in any case everything depends on theories of meaning. Much the same applies to the philosophy of mathematics, though for different reasons which can here be set on one side.

Certainly, we cannot extrapolate from the philosophies of morals and of mathematics to the philosophies of science or of religion. Michael Devitt, indeed, urges with special reference to the philosophy of science that we must not 'put the semantic cart before the realist horse',[5] realism being the doctrine that 'material objects exist externally to us and independently of our sense experience';[6] and this definition can be construed as the sectoral equivalent of Warner's general account presented above. However, when we come to the philosophy of religion, we have to avoid the associated temptations of representing all non-believers as anti-realists and all believers as realists. Indeed, as Peter Byrne remarks,

> while Richard Swinburne and Quentin Smith disagree over whether God is real, they agree in a realist interpretation of theistic discourse. Contrariwise, while Smith and [Don] Cupitt both agree that the God of traditional theism does not exist, the former gives a realist interpretation of theism while the latter does not. What Smith and Swinburne agree on is that the governing intent of theistic discourse and belief is a realist one. Realism brings out the meaning of theism.[7]

Byrne's classifications seem correct, and it is accordingly difficult to deny that realism about religion concerns issues of interpretation and meaning. And this recognition in turn suggests that Dummett's semantic account of realism and anti-realism warrants serious consideration, whether we agree with him about theories of meaning being philosophically fundamental, or with Devitt that this can sometimes put the cart before the horse. This approach has the advantage of treating anti-realism in several fields together, rather than religious anti-realism in isolation.

The roots of anti-realism are to be found in more places than one. Hume's scepticism about substance, about causation, about identity and about reasoning by induction suggested that human beings lack knowledge of the world around

---

4 J.L. Mackie (1977), *Ethics: Inventing Right and Wrong*.

5 Michael Devitt (1984), *Realism and Truth*, p. 2.

6 Devitt, ibid., p. 11. Devitt, who endorses this definition, is here quoting R.J. Hirst, 'Realism' in Paul Edwards (ed.) (1977), *The Encyclopedia of Philosophy*, 77–83.

7 Peter Byrne (2003), *God and Realism*, p. 4.

them (although he saw fit to ignore such problems for practical purposes), and also that human claims have to be geared to the limits of human faculties, a stance also advanced, albeit on different grounds, by Comte.[8] Among verificationists of the pragmatist variety, the move has often been made of gearing interpretations of reality and of truth to the limits of human faculties and of human inquiry,[9] a move they could claim to find in the work of some of Kant's idealist followers. David Conway, whose survey of the history and sociology of anti-realism I can commend to readers, rightly identifies Nietzsche as another significant source.[10] The understanding of objectivity on the part of the later Wittgenstein as grounded in the shared concepts of a community can be (and, as we shall see, has been) given an anti-realist interpretation.

A further significant source has been verificationism of the logical positivist variety. While several of the leading positivists were realists, others, like Rudolph Carnap and Otto Neurath, rejected the correspondence theory of truth and adopted understandings of truth and of justification related to human limitations.[11] In matters of theistic discourse, one verificationist and anti-realist interpretation has been mentioned above, that of Braithwaite.[12] R.M. Hare's interpretation of claims about God's love as expressions of fundamental commitment (or 'Bliks') presents a different form of anti-realism, and was also influenced by empiricist distinctions between the realm of the observable and the definable (in his view the realm of statements) and contrasting uses of language, as in ethics and religion.[13]

The anti-realism of Michael Dummett, however, is a form of verificationism that is far removed from such non-cognitivist views. Yet Dummett's anti-realist arguments are distinctive, and have been widely debated by philosophers who would not dream of considering themselves verificationists. These arguments are as important as the truth of anti-realism would be, not least because of its possible implications for religion; accordingly they are discussed in the coming section.

Section 2: Dummett's Anti-realist Arguments

Dummett depicts realism and anti-realism as follows:

---

8 C.J. Misak, *Verificationism*, pp. 8–25.

9 Misak, pp. 97–127, 163–71.

10 David Conway (2000), *The Rediscovery of Wisdom*, pp. 6–12.

11 Misak, pp. 94–6. On the debate between Neurath and Moritz Schlick, see Israel Scheffler (1996), 'Epistemology of Objectivity', in Peter J. McCormick (ed.), *Starmaking: Realism, Anti-Realism and Irrealism*, 29–58.

12 R.B. Braithwaite (1955), *An Empiricist's View of the Nature of Religious Belief*.

13 R.M. Hare, 'Theology and Falsification', in Flew and McIntyre, *New Essays*, 99–103.

> Realism I characterise as the belief that statements of the disputed class possess an objective truth-value, independently of our means of knowing it; they are true or false in virtue of a reality existing independently of us. The anti-realist opposes to this the view that statements of the disputed class are to be understood only by reference to the sort of things which we count as evidence for a statement of that class ... The dispute thus concerns the notion of truth appropriate for statements of the disputed class; and this means that it is a dispute concerning the kind of *meaning* which those statements have.[14]

Dummett is not suggesting (as some of the logical positivists did) that sentences expressing statements of the disputed class are meaningless, much less statements about God, but rather that their meaning must be construed as subject to a constrained, anti-realist notion of truth, for which these same statements are neither true nor false. Examples of the disputed class include, according to Crispin Wright, 'unrestricted spatial or temporal generalisations, many subjunctive conditionals, descriptions of the remote past, hypotheses about the mental life of others or of animals';[15] the anti-realist raises questions about our grasp as well as about our possible knowledge of statements of these kinds.

While Dummett did not include (and would almost certainly not have wished to include) sentences or statements about God, there is nothing to prevent religious anti-realists from adding such language to the list, for example: 'All living organisms are creatures of God' and 'The Earth is the Lord's'. Could these claims lie beyond our possible knowledge, or even beyond our grasp? Or could it be that realist interpretations make them unacceptable or even incomprehensible? Such questions involve reflection on the case for generic anti-realism, considered in this and the coming section, which is crucial in any case for the philosophy of meaning and for metaphysics.

Dummett advances at least two arguments against generic realism and in support of generic anti-realism, which are better regarded as challenges to reconcile claims to which the realist is held to be committed. In the current section, these key challenges are appraised. How secure is Dummett's case for anti-realism?

The Acquisition Challenge rests, to follow Hale's account, on 'our training in the use of language' consisting in our being taught to accept statements as true in one range of circumstances, and to reject them in a contrasting range of circumstances, and goes on to ask how we can 'come to know what it is for a statement to be true, or false, in virtue of' a state of affairs which, if it obtains, does so 'undetectably'. The realist seems hard-pressed to explain how we can assign as integral to the meaning of statements truth-conditions involving states

---

14 Michael Dummett [1963], 'Realism', reprinted in Dummett (1978), *Truth and Other Enigmas*, 145–65, p. 146.

15 Wright (1987), *Realism, Meaning and Truth*, p. 53.

of affairs that can play 'no part in the process by which the meanings of those statements are learned'.[16]

The Manifestation Challenge proceeds in a parallel manner; again I follow Hale's résumé. 'If the meaning of a statement consists in its having certain … truth-conditions' which in disputed-class cases may be evidence-transcendent, then understanding it, or knowing its meaning, involves possessing knowledge (whether explicit or implicit) of such truth-conditions. But knowledge of meaning cannot invariably be explicit knowledge, and generally consists in 'knowledge of how to use' the relevant sentence, and thus in having 'certain practical abilities'. The realist now seems hard-pressed to explain by what practical abilities a grasp of 'evidence-transcendent truth-conditions' could be manifested.[17]

A realist could reply to these arguments that knowledge of a declarative sentence's meaning is not in general equivalent to knowledge of its truth-conditions, since the truth-conditions of the very many sentences that involve referring expressions depend on which reference is being made, and this in turn depends on the context, and not only on the meaning of the sentence. (The same example of a sentence as used in the previous chapter, 'She hit him with it', may suffice to make this clear.) Clearly the above Challenges need either to exclude such sentences, or to be modified to accommodate them. However, for many other sentences there is no such problem, since one and the same statement is expressed whenever and wherever they are used. Hence I will not rely on this objection, and will construe the Challenges as concerning those sentences where it makes no difference. These will also be cases where 'the meaning of the sentence' and 'the meaning of the statement expressed' are effectively identical.

However, a response to the Acquisition Challenge has been presented by Wright and endorsed by Hale. This Challenge effectively relies on our needing training in a sentence's field (whether spatial, temporal, historical, psychological or hypothetical) in order to grasp its meaning. But our understanding of most sentences, the realist can retort, comes not from direct links between the sentences and recognizable corresponding states of affairs, but from prior grasp of their component expressions plus a grasp of how to combine them in semantically significant ways. Thus until it can be shown that we cannot grasp realist truth-conditions in this way, there is nothing to prevent the realist claiming that we grasp them by learning sentences which have these truth-conditions, and we learn these sentences by understanding their component words and syntax. This response does not of itself vindicate realist claims, but in the absence of an argument to show that we cannot grasp realist truth-conditions in the first place it is a sufficient response to meet the Acquisition Challenge.[18] Indeed, a matching sectoral response is open to a realist concerning language about God and creation.

---

[16] Bob Hale (1997), 'Realism and Its Oppositions', in Bob Hale and Crispin Wright (eds), *A Companion to the Philosophy of Language*, 271–308, p. 275.

[17] Hale, ibid., pp. 275–6.

[18] Wright, ibid., pp. 15–16; Hale, ibid., pp. 279–80.

Where Dummett's arguments are concerned, then, much turns on the Manifestation Challenge. Alexander Miller here distinguishes a strong version of this Challenge and a weak version. The strong version seeks to establish the falsity of semantic realism. But when defenders of this doctrine such as Peter Strawson reply that relevant abilities can be displayed in a variety of responses to recognizable conditions, abilities such as appraising evidence and drawing implications, and that realism can thus accommodate the request to identify practical abilities mentioned in the Challenge,[19] Hale's response amounts to the claim that Strawson's account gives no reason to accept semantic realism in preference to anti-realism, and not that semantic realism is false.[20] But this now embodies what Miller calls 'the weak version' of the Challenge, which seeks to establish the weaker conclusion 'that we cannot justify accepting semantic realism *solely* on the basis of considering the practical abilities which constitute linguistic understanding'.[21] The switch to the weak version already embodies a major concession, at least if we begin by endorsing Wright's view (despite its tension with other claims of his) that 'realism simply supplies the most natural pre-philosophically agreeable' theory in analytic metaphysics.[22]

But in any case the weak version turns out to be inconclusive. Thus if we grant, as Wright does, that there is a platitudinous, truistic connection between the notion of the content of an assertion and the notion of its truth-conditions,[23] then we can reason validly from this and the uncontested fact that we are capable of understanding undecidable assertions such as 'Caesar sneezed 15 times on his 19th birthday' to the conclusion that our understanding 'consists in a grasp of truth-conditions which, if they obtain, we may be incapable, even in principle, of detecting'.[24] In this way the realist can show how the abilities of a competent speaker comprise grasp of truth-conditions which are potentially recognition-transcendent, which is just what the weak version of the Manifestation Challenge demands.[25] The only apparent way for the anti-realist to challenge this argument is to interpret semantic realism as holding that speakers can understand not just the truth-conditions of some declarative sentence which may in fact obtain undetectably, but also what it would be for the truth-conditions of such a sentence to obtain undetectably, and in some unknown way beyond their recognition.[26] But this involves building into the Challenge the requirement that the semantic

---

19 Peter Strawson (1977), 'Scruton and Wright on Anti-Realism', *Proceedings of the Aristotelian Society*, **77**, 15–22, p. 16.

20 Hale, ibid., p. 281.

21 Alexander Miller (2002), 'What is the Manifestation Argument?', *Pacific Philosophical Quarterly*, **83**, 352–83, p. 360.

22 Wright, ibid., p. 1; for this appraisal, see Miller, ibid., p. 361.

23 Wright (1992), *Truth and Objectivity*, pp. 23–4.

24 Miller, ibid., p. 367.

25 Miller, ibid., p. 368.

26 Miller, ibid., p. 369.

realist must show not only 'that the understanding of competent speakers is as realism would describe it' (a reasonable enough request), but also that it involves *awareness that the relevant truth-conditions hold undetectably* and thus 'manifests an understanding of realism'.[27] But the latter is an unreasonable demand; realists cannot be expected to show that their theory of understanding requires speakers to grasp a metaphysical theory.

It might be objected that Miller's example of an undecidable sentence includes referring expressions, and thus falls outside the scope of the Manifestation Challenge as interpreted here. But fortunately there is no ambivalence about these references; hence there is no objection to it being treated as a relevant example. Accordingly I can endorse Miller's conclusion that the Manifestation Challenge fails in both its versions.

The Manifestation Challenge, it may be added, effectively relies on our needing to be able to use a sentence in order to understand it. But this requirement is unduly demanding; for example, I can grasp what Latin sentences meant among Cicero's contemporaries without having the ability to use them in spoken communication. There again, a deaf person could understand a spoken warning through lip-reading, but might be unable to utter such a warning, and might imaginably not be proficient enough at sign-language to convey such a warning at all. Thus the assumption of the Challenge that understanding involves use in the sense of the abilities of a competent speaker is misguided. There is much more communication than it suggests, much of it to recipients who are, for relevant purposes, non-speakers. The above discussion effectively relies, perhaps reasonably, on the assumption that the generality of those who understand language are competent speakers of a relevant language. But in fact the phenomenon of linguistic understanding could remain as widespread as at present, or virtually so, even if some cosmic catastrophe brought it about that a majority of speakers were struck dumb, for those newly rendered speechless could continue to understand the talk of surviving speakers (plus the multiplicity of public signs that have almost become co-extensive with the distribution of human society itself).

But if this objection stood alone, then anti-realists could deploy a Recognition Challenge, mirroring the Manifestation Challenge discussed above, but concerned with the abilities required for recognition of declarative sentences rather than for their use. Christopher Norris has in fact discovered such an argument for capacities of recognition as crucial for understanding (albeit without an explicit Recognition Challenge to realists) in the work of Dummett;[28] other commentators seem to have neglected this argument. However, an exactly parallel set of replies could

---

[27] Miller, ibid., p. 371.

[28] Christopher Norris (2004), 'Staying for an Answer', *Philosophy and Social Criticism* **30** (7), 777–98, p. 785; Dummett (1976), 'What is a Theory of Meaning? (II)', in Gareth Evans and John McDowell (eds), *Truth and Meaning: Essays in Semantics*, 67–137, pp. 110–11.

be advanced to those of Miller, showing that this possible Challenge also proves futile. Dummett seems therefore not to make out his case for anti-realism.

Section 3: Further Arguments for and against Anti-realism

Crispin Wright advances two further arguments against realism, one modelled on Dummett's Challenges and one on the work of Ludwig Wittgenstein. It is argued here that these further arguments are also inconclusive. While this is a matter of considerable importance both for semantics and for metaphysics, readers eager to move without delay to applications of anti-realism to God and creation could turn to the following three sections.

The argument that Wright presents as parallel to the Acquisition and Manifestation Arguments is the Argument from Normativity. Since meaning is normative, knowing the meaning of an expression involves knowing 'a set of constraints to which correct usage must conform'. So 'to give the meaning of an expression is to describe such constraints; nothing has a claim to be regarded as an account of a statement's meaning which does not succeed in doing so'. But the realist's conception of truth-conditions can make no such claim. To be able to appraise satisfactorily a statement which is true but may be undetectably so, a person needs to grasp the possibility that no evidence in its favour is available, so as to be able to distinguish between the truth-conditions of the statement itself, those of the statement that it is undetectably false, and those of the statement that there is no available evidence bearing on its truth or falsity. But the realist cannot bring this off, claims Wright.[29] When truth is realistically conceived, the truth-conditional conception of meaning is thus in tension with the normativity of meaning, and the tension can only be resolved by adoption of an anti-realist conception of truth.[30]

Here we can again use Miller's example, 'Caesar sneezed 15 times on his 19th birthday'. It can be granted to Wright that those who understand this claim must be able to distinguish its meaning from that of 'Caesar sneezed 15 times on his 19th birthday, but there is no available evidence for believing this' and from 'There is no available evidence that Caesar sneezed 15 times on his 19th birthday'. But this they could do through exercising the abilities noted by Strawson (abilities that pose no problems for realists), such as the ability to draw relevant implications and to assess the related evidence (or lack of evidence). Thus the third of these possible claims relates to lack of evidence available in the present, a claim for which there could in fact be grounds such as the silence of sources like Suetonius, our regular source for the life of Caesar; while the first could possibly be advanced on the basis that most people sneeze at least 15 times per day, and that active people such as Caesar sneeze more than others (and so it is likely that Caesar sneezed as much as

29 Wright, *Realism, Meaning and Truth*, p. 24.

30 Wright, ibid., pp. 24–6.

this), a basis that could be held not to be verifiable (depending on how strict the requirements for verification are taken to be). In such ways, the realist can claim, the requirements attaching to the normativity of meaning are satisfied.

Besides, Wright seems to require too much when he requires that a realist account of understanding the claim about Caesar involves those concerned understanding that it is or may be undetectably true (or therefore, as he goes on to suggest, understanding what it is to seek for undetectable truth).[31] Many claims and conjectures are made or given consideration without awareness of what would be required to vindicate them, and often of whether their vindication is possible. Indeed Fermat's last theorem used to be held unprovable, but has recently been proved by Andrew Wiles; and evidence of the continuing existence of the coelocamph used to be held to be unavailable (when this species of antique fish was considered extinct) but it now widely acknowledged. Thus to require grasp of a statement's actual or possible epistemological status as a condition of understanding that statement seems overdemanding, both on the above grounds and also because grasping a claim's epistemological status in any case presupposes grasping the claim itself, a feat which would therefore have to be possible independently. Indeed Wright here seems to be requiring the realist account of those who grasp or use ordinary language to include a grasp of the epistemological status of relevant claims well beyond what can reasonably be expected.

Wright's final argument is based on what interpreters of Wittgenstein refer to as 'Wittgenstein's rule-following considerations'. For discourse to have objectivity of truth in the manner claimed by realists, it must also embody an objectivity of meaning which is independent of our opinions and judgements of the matter in question. But (in Miller's words) 'Wright argues that we must view meaning as judgement-dependent if we are to avoid the sceptical paradox about meaning which [Saul] Kripke reads into Wittgenstein's rule-following considerations'.[32] While the debate about these considerations cannot be discussed in any detail here, Wright argues that a person's self-ascriptions of intention (and thus their claims about what their own meaning is or was) are normally authoritative (or decisive) and thus determine their meaning.[33] Hence their meaning depends on their judgements, at least in cases of this kind. Thus meaning is judgement-dependent (at least sometimes), and the realist position about both meaning and truth collapses.

John McDowell, however, disputes this view both as a cogent interpretation of Wittgenstein and as a plausible theory. Wittgenstein probably did not accept the sceptical paradox that there are no such things as facts about meaning;[34] and

---

[31] Wright, ibid., p. 26.

[32] Miller (1998), *Philosophy of Language*, p. 291.

[33] Miller, ibid., pp. 205–7.

[34] John McDowell (1992), 'Meaning and Intention in Wittgenstein's Later Philosophy', *Midwest Studies in Philosophy*, **17**, 40–52; Miller, ibid., p. 215–16. For another view, see Bob Hale, 'Rule-Following, Objectivity and Meaning', in Hale and Wright, 369–96, pp. 380–91.

Wright is misguided in denying that meaning and intending can be a matter of having something in mind. A meaning or intention can 'determine in advance' (independently of subsequent judgements) 'what patterns of behaviour accord with it and which do not',[35] and subsequent judgements presuppose such facts about meaning, rather than determining its content.[36] Further, if we abandon 'the judgement-independent conception of meaning, we will not be able to find room for the idea that meaning is genuinely normative'.[37]

So Wright's anti-realist case, rather than being clinched on the basis of this argument, threatens to undermine the very basis (that of the normativity of meaning) on which his previous argument was grounded. There must also be doubts about any theory that questions the objectivity of meaning, for any such theory has the potential to destabilize the meaning of the sentences in which it is expressed, and thus to undermine itself.

Thus none of the standard arguments challenging semantic realism is conclusive. Besides, these arguments all rely on what Miller calls 'the Truth-Conditional Conception of Meaning and Understanding' (TCCMU), namely that 'a sentence having the meaning it has consists in its having a certain truth-condition, and that a speaker's understanding that sentence … consists in his having grasped the relevant truth-condition'.[38] Indeed in the absence of this conception, a verification-transcendent conception of truth could readily be combined with acceptance that understanding consists in a complex of practical abilities, no longer open to challenge about the truth-conditions that TCCMU requires to be understood. However, grounds exist for doubting the TCCMU.

The clearest argument against TCCMU canvassed by Miller relates to the causal theory of the meaning of natural kind terms and what is involved in understanding sentences involving such terms. Since on this theory those who use or understand such sentences need not be aware of the causal relationships that facilitate the meaning of what they understand, they cannot be expected to know which the relevant truth-conditions may happen to be.[39] But since the causal theory is itself controversial, I will rest nothing on this argument.

A second argument has already been presented in the previous chapter. Sentences involving referring expressions can be used to express unlimited numbers of propositions, each with distinctive truth-conditions. Hence the meanings of sentences cannot be mapped against 'their' truth-conditions, for the truth-conditions of a single sentence can be almost infinitely diverse. This objection, however, does not apply to theories concerning the meaning of utterances of

---

35 Miller, ibid., pp. 216–17.

36 Paul Boghossian (1989), 'The Rule-following Considerations', *Mind*, **98**, 507–49, p. 547.

37 Miller (who is here interpreting McDowell), ibid., p. 292.

38 Miller (forthcoming), 'Realism and Antirealism', in Lepore and Smith (eds), *Oxford Handbook on the Philosophy of Language*, Section 4.

39 Miller, ibid., Section 6.

a sentence or occasions of use or of understanding, since those occasions, in conjunction with linguistic conventions, generally facilitate the identification of the proposition expressed, and hence of its truth-conditions. TCCMU could be revised so as to concern the meaning and understanding of utterances (etc.), and thus immunized against this argument. (But since the truth-conditions could only be identified through the meaning, they could not in any case be regarded as the basis of the meaning.[40])

A third argument arises from the analogical use of language that was discussed in Chapter One. When terms like 'intelligent' are used of frogs or of computers, let alone of deities, they have different synonyms and antonyms from those associated with their use of human beings, and the related truth-conditions will be correspondingly different. Since there are unlimited possibilities for analogical uses of language, this makes it prohibitively difficult to specify the truth-conditions of any expression amenable to analogical use in advance of knowledge of that use.[41] Once again, TCCMU turns out to have limited applicability, at best. But the revision just suggested, making it concern the meaning and understanding of utterances (etc.), could well remedy this problem, since awareness of the occasion of utterance would normally convey understanding of which way, if any, an expression is being used analogously, and in principle allow relevant truth-conditions (or at least their range) to be discovered. However, where language is used metaphorically, it is less clear that there is any way to rescue a truth-conditional theory of meaning; hence TCCMU should best be restricted to literal and analogous uses of language, with metaphorical language recognized as beyond its scope. (And the onus would be on the defender of TCCMU to establish case by case that the uses of language to which it was to be applied were non-metaphorical: no easy task.)

The fourth argument, however, should be taken more seriously, and was raised in the later work of A.J. Ayer, reflecting on the work of Quine. 'The truth-conditions of a given statement', he remarks, 'cannot be separated from those of the other empirical constituents of the scientific corpus of which it is an ingredient.' Given a holistic view of meaning (however modest), the meanings of single sentences are likely to be intertwined, and their truth-conditions also. Further, adds Ayer, 'the truth-conditions of the whole corpus are not fully specifiable'.[42] Thus, if we adopt even a weak holistic theory (a stance that is far from absurd), no truth-conditional theory of meaning is likely to yield a clear and satisfactory account of specific truth-conditions; and the scope of this conclusion extends by parity not only to the corpus of science but to other

---

[40] Paul Horwich (1998), *Meaning*, pp. 72–4.

[41] Martin Warner (1992), 'Language, Interpretation and Worship – I', in Warner (ed.), *Religion and Philosophy*, 91–108, pp. 102–3.

[42] A.J. Ayer, 'Reply to Michael Dummett', p. 153. Dummett, however, has reservations about holistic theories of meaning; see Dummett (1987), 'Reply to Brian Loar', in B. Taylor (ed.), *Michael Dummett: Contributions to Philosophy*, 269–80, p. 273.

regions of discourse as well. This does not show that TCCMU is false, but suggests that even if it is true, we cannot know this. Ironically, this suggests that only realists could consistently appeal to it; anti-realists could not, since it comprises a verification-transcendent claim.

Hence anti-realists are in no position to assume the truth of TCCMU, and either convict realists of inconsistency on that basis, or devise challenges on the assumption that they cannot as much as question it. Meanwhile realists can maintain that even if the above responses to the various challenges and arguments fail, they can retain adherence to a verification-transcendent conception of truth and related understandings of meaning simply through the expedient of disowning TCCMU. This granted, realism remains undisturbed in its position as 'the most natural theory' of metaphysics to adopt.[43]

Section 4: Rorty's Anti-realism

A more sweeping form of anti-realism than those of Dummett or Wright or Hale has been advocated by Richard Rorty, for whom realism depends on an outmoded belief in creation. For Rorty, truth is a property not of states of a world independent of the human mind – he writes here as if realists said otherwise – but simply of sentences; and sentences are elements of human languages, which are human creations. Nothing but descriptions can be true or false, and descriptions, unlike the world, are dependent on human beings. The suggestion that truth belongs to the world or to its 'intrinsic nature', Rorty maintains, 'is a legacy of an age in which the world was seen as the creation of a being who had a language of his own'.[44] With this suggestion belongs 'the claim that the world splits itself up, on its own initiative, into sentence-shaped chunks called "facts"'. But this makes it easy to treat the truth as identical 'either with God or with the world as God's project'.[45]

Thus Rorty is launching a global assault not only on the correspondence theory of truth, but also on realism as a theory of meaning in which language is intended to depict a reality independent of ourselves. Further, he claims that these conceptions are implications of belief in creation, and are unmotivated without that belief. It could here be remarked that the eighteenth-century philosopher George Berkeley once maintained that the regular world around us should be understood as an expression of God's language,[46] but felt obliged to say this only because he rejected belief in material objects; his contemporaries were able to disown this view because of their shared conviction that material objects are

[43] Wright, *Realism, Meaning and Truth*, p. 1.

[44] Richard Rorty (1989), *Contingency, Irony and Solidarity*, p. 5.

[45] Rorty, ibid., p. 5.

[46] George Berkeley [1732] (1950), *Alciphron, or the Minute Philosopher*, in George Berkeley (1948–57), *Works*, ed. A.A. Luce and T. Jessop, vol. III.

real whether God is their creator or not. Besides, as Roger Trigg has said in reply to Rorty, the Enlightenment was often self-consciously atheistic, but is not for that reason to be regarded as irrational, let alone inconsistent.[47] Meanwhile the undoubted historical and logical links between belief in creation and the rise of modern natural science (see Chapter One above) lend no support to Rorty's anti-realism. They suggest that the natural world is suited to human discovery, but not to human creation; much as we may need humanly devised conceptual schemes and theories, nature is not conjured into existence by us, but given.

Rorty holds that the only way to argue his case against the givenness of the world is to 'exhibit the sterility of attempts to give a sense to phrases like "the way the world is" or "fitting the facts"'.[48] Further attempts to argue the case (beyond his argument about the dependence of truth on human decisions) commit us to vocabularies embodying the old order that he is opposing, and reason is in general a function of the languages we happen to speak. But no vocabularies are better representations of the world than others;[49] to suppose otherwise, Rorty claims, is to have in mind a picture of the universe as created by a person with a language and a plan of his own, and the only way to overthrow this supposition is to 'de-divinize the world'.[50]

Rorty's interesting suggestion that realism entails theism has already been addressed. Rebutting his conceptual relativism, however, involves contesting his argument for the dependence of truth on human decisions, and drawing to attention why its conclusions are unacceptable. One way of expressing what is wrong with the argument (beyond remarking that it is not strictly sentences that are true or false in the first place) is to highlight its confusion of what Peter Byrne calls 'truth-bearers' and 'truth-makers'.[51] Granting that truth is a property of units of language (truth-bearers) does not show that its content is determined by the individual or communal wordsmiths who devise those units; sentences may be necessary for truth, but they are not sufficient. To cite Byrne's example, the North Downs are not dependent on human representations, nor are most facts about them (even though some result from human activity). Thus the Downs would continue to exist, with many of their properties intact, if some catastrophe were to extinguish humanity as a whole.[52] But it is nothing but things having the properties that they do that constitutes the states of affairs that are truth-makers, and that make units of language (or clusters of such units) true or false. (That is why scientists and historians, for example, seek to make their descriptions

---

47 Roger Trigg (1992), 'Reason and Faith – II', in Warner, *Religion and Philosophy*, 33–43, p. 37.

48 Rorty, ibid., p. 20.

49 Rorty, ibid., p. 21.

50 Rorty, ibid., p. 21.

51 Byrne, *God and Realism*, p. 53.

52 Byrne, ibid., pp. 21–2.

conform to facts, rather than to create them.) This thought can be re-expressed in J.L. Austin's classic presentation of the correspondence theory of truth:

> A statement is said to be true when the historic state of affairs to which it is correlated by the demonstrative conventions (the one to which it 'refers') is of a type with which the sentence used in making it is correlated by the descriptive conventions.[53]

Austin of course appeals to semantic conventions in force within the speaker's or writer's linguistic community, but this gives no foothold for any conventionalist theory of truth. However, there is no need to endorse Austin's specific theory of truth to see that truths are dependent on the way things are, and not just human creations as Rorty suggests.

Rorty would no doubt reject this argument as being expressed in an unsatisfactory vocabulary. However, this objection raises the problem for him of supplying an impartial, non-question-begging criterion for a satisfactory vocabulary. In any case there is one further kind of objection which the conceptual relativist cannot evade. If, as Rorty suggests, different philosophical accounts are merely different narratives, none rationally preferable to the others except from the perspective of their cultural function, then there is no rational basis for the acceptance of his own account either. Further, if, as his claims suggest, his own theory is just a narrative that reflects no facts not subject to human decision, then there could be no grounds (beyond its utility) for preferring it to a contrary theory. Thus Rorty supplies no grounds for global anti-realism beyond the grounds rejected in previous sections, nor for the application of this anti-realism to God and creation. He also makes claims that appear to conflict with everyday beliefs about states of affairs, the independence of most of them from human representations, and thus their givenness. Indeed Rorty's claims and arguments do nothing to impede arguments from the very contingency that he stresses – to creation.

Section 5: Cupitt's Anti-realist Faith

Referring to Rorty's *Contingency, Irony and Solidarity*, the Anglican priest Don Cupitt wrote as follows:

> I agree with Rorty that we … now find ourselves having to live without old-style metaphysical or theological underpinning for our final vocabulary. Yes: for me too everything is contingent, a product of history and open to re-assessment.[54]

---

53 J.L. Austin (1970), *Philosophical Papers*, ed. J.O. Urmson and G.J. Warnock (2nd edn), p. 122.

54 Don Cupitt (1990), *Creation out of Nothing*, pp. 15–16.

The contingency of our language and of our perspective supposedly removes any possibility of making credible timeless claims. Indeed Cupitt holds, with Rorty, that 'language creates reality', and that accordingly 'God, like us, is made only of words'. Shifts of language accompany shifts of perspective, and Nietzsche's view is adopted that all truths are relative to perspective; hence the suggestion that God is shaped (indeed created) by human language. The view that language seeks to depict a reality independent of our conceptions is simultaneously discarded. As for creation, 'We can no longer distinguish clearly between the sense in which God creates, the sense in which language does, and the sense in which we do'; for creation is not the work of God, but of human beings.[55] As Martin Warner aptly comments,

> There is a clear break here with the traditional religious belief that we are created and sustained by God rather than the reverse, and that consequently God's choice is prior to that of his creatures and human creativity dependent on the creator.[56]

In an eloquent and evocative survey of the history of philosophy, Cupitt represents the growing recognition of such perspectivism as inevitable.[57] At first, the far-reaching implications of the discovery two hundred years ago that human interpretations shape all thought and understanding were played down in different ways by thinkers who purported to discover necessities of different kinds, either in thought or in history (thinkers as diverse as Kant, Hegel and Marx). But with Nietzsche such alleged necessities were rejected, and the term 'anti-realism' was eventually coined to express the Nietzschian slogan 'There are no facts, only interpretations'. It was recognized, claims Cupitt, that there is no knowledge independent of perspectives, of which there are indefinitely many; and none of them can claim certainly or unchanging finality. This new critical thinking became clearer with its application to language in Peirce's pragmatism and in the work of the later Wittgenstein. Meanings ceased to be regarded as spirit-like entities (however impersonal) and were understood instead as uses and functions that evolve in the course of history. There are no unchanging meanings, nor, by the same token, any timeless truths. 'The whole world of meaning, which is the true starting point for philosophy, is by its very nature shifting all the time like the prices in a stock market, as human power relations shift'.[58]

Cupitt introduces here an instructive example:

> I personally am prepared to fight tooth and nail for modern evolutionary biology against creationism. But I cannot claim that current evolutionary theory is, in any part of it, objectively, dogmatically and perennially just true. On the contrary, over

---

55 Cupitt, ibid., pp. ix–x, 97.

56 Martin Warner (1992), 'Introduction', in Warner, *Religion and Philosophy*, p. 5.

57 Cupitt (1993), 'Anti-Realist Faith', in Joseph Runzo (ed.), *Is God Real?*, 45–55, pp. 45–51.

58 Cupitt, ibid., p. 47.

> the generations to come I expect that every bit of current evolutionary theory will be replaced by something different. In this shifting relativistic world of ours, we can still choose our values and fight for them, but our beliefs won't have the old kind of permanent anchorage in an unchanging ideal order.[59]

His philosophical and theological symposiasts are likely to have felt more than a glimmer of sympathy with this rejection of creationism and highly qualified espousal of evolutionary theory (see Chapters Four and Six below), although the suggestions that we choose our values (rather than recognize or discover them) and that nothing is true both yesterday, today and tomorrow clearly gained less acceptance.[60] The view that scientific claims are subject to constant revision, after all, can be maintained without accepting that there are no unchanging truths, for example in logic or in mathematics. It is also fully compatible with a realist view of the philosophy of science, and with recognition that the world has a character independent of our language and conceptions, and that revisions of scientific theory enable us to describe it more adequately.[61] Besides, whatever one's perspective, current meanings clearly generate conceptually necessary truths, true wherever those meanings remain current. For this reason, necessary truths actually remain the framework of each and every perspective.

Cupitt regards the emergence of anti-realism as a liberation for religion. Christianity is liberated from 'the traditional outlook of Christian Platonism',[62] comes to accept that religion belongs wholly inside the one common world of constantly changing human culture and history, which has no outside,[63] is stripped of its essentialist claims to timeless truth,[64] but is all the more autonomous in its advocacy of sacrificial love.[65] It is thus human believers who are freed to be creative, in place of God, who alone used to be so regarded.[66] Indeed the idea of God's speaking words that belong to no language community and yet have meaning for Him is held up to ridicule.[67] While the scope of Cupitt's anti-realism is global, the realist Christianity of earlier ages is, for him, understandable; but it was usually qualified by a recognition of the inadequacy of human words to describe

---

59 Cupitt, ibid., p. 48.

60 See Brian Hebblethwaite, 'A Critique of Don Cupitt's Christian Buddhism', in Runzo, *Is God Real?*, 135–48.

61 For defences of critical realism in the philosophy of science, see Michael C. Banner (1990), *The Justification of Science and the Rationality of Religious Belief*, pp. 35–7; W.H. Newton-Smith (1981), *The Rationality of Science*; Roy Bhaskar (1986), *Scientific Realism and Human Emancipation*.

62 Cupitt, 'Anti-Realist Faith', p. 49; *Creation Out of Nothing*, pp. 84–90.

63 Cupitt, 'Anti-Realist Faith', p. 50.

64 Cupitt, ibid., p. 53.

65 Cupitt, ibid., 54–5; Cupitt (1986), *Life Lines*, p. 130.

66 Cupitt, 'Anti-Realist Faith', p. 55.

67 Cupitt, ibid., p. 51.

the divine.[68] Such negative theology paved the way for the non-supernaturalist, anti-realist faith of today.

However, as Cupitt would have to agree, there are other ways of reading history. Thus the recognition of the apparent inadequacy of human language about God generated theories of analogy (see Chapter One above) which show how sense can be made of such language despite the problems when finite believers talk of an infinite God. Further, as Hebblethwaite remarks, it is not mandatory to endorse the idealist elements in Kantian thought, much less those of his successors;[69] indeed the very language of interpretations and of what appears to us to be the case is parasitic on the language of what is actually the case. Again, the discovery of the romantic period that each historical text emerges from a specific historical context has valuably contributed to critical scholarship, but does not imply that the content of texts can never aspire to be true in a manner that transcends both language and contexts.

This brings us back to the core issue of Cupitt's perspectivism, the claim that all truth is relative to perspective, whether that of a community or culture or of an individual. But this claim undermines itself, for its truth must be also be relative, relative to perspectives that endorse perspectivism; and it is contrary to perspectivism to acknowledge truths that bestride or transcend perspectives. So either it applies to itself and is self-stultifying, as Stephen T. Davis suggests,[70] or there are exceptions to its scope which allow of reasoning between and across perspectives, in which case there is room for the kind of reason which perspectivism rejects. But if so, claims need not always be relative to perspective, and there may remain scope for the kind of 'rational realism' which asserts that we sometimes have good reasons for accepting the approximate truth of scientific theories,[71] and similarly for the kind of religious faith which claims to have good reasons for belief in its transcendent object, and does not assimilate or reduce belief, in anti-realist fashion, to commitment to the values of that faith.[72] Such belief could even supply grounds for the related religious commitment, a possibility unavailable to anti-realist faith. For all that Cupitt's anti-realism shows to the contrary, it could concern a God capable of speaking meaningfully to each and every linguistic community, and even of calling such communities into existence.

Section 6: Phillips's Religious Projectionism

D.Z. Phillips suggests that we should follow Wittgenstein in rejecting both realism and non-realism. For while realists misconstrue religious language, non-realists

---

[68] Cupitt, ibid., p. 52.

[69] Hebblethwaite, p. 147.

[70] Stephen T. Davis, 'Against "Anti-Religious Faith"', in Runzo (ed.), *Is God Real*, 57–9, p. 58.

[71] Banner, op. cit., p. 35.

[72] Trigg, 'Reason and Faith – II', pp. 33–5, 42–3.

(including Cupitt, seen from Phillips's perspective) mistakenly allow that realism conveys 'what faith once meant for people, but argue that this conception of faith cannot be sustained today'. By contrast, followers of the Wittgensteinian critique regard both realism and non-realism as equally confused; for the religious thought and practice of all generations is to be understood on a non-realist basis as it is, without any need for revisionism such as that of Braithwaite or Hare or Cupitt.[73]

In *The Concept of Prayer* and *Faith and Philosophical Enquiry*, Phillips appeals to Wittgenstein's *Lectures and Conversations on Aesthetics, Psychology and Religious Belief* and claims that Wittgenstein shows that two people one of whom believes in God while the other does not are not contradicting one another. 'The main reason for the difference is that God's reality is not one of a kind; He is not a being among other beings. The word "God" is not the name of a thing'.[74] For it does not make sense to speak of God ceasing to exist, or to ask what brought Him into existence; we have to take into account what makes sense within the relevant language-game. That God is not an object among other objects, and not subject to many of the questions that apply to the denizens of space and time, is likely to be agreed on all sides; but when Phillips effectively claims that God is not an individual at all, and thus not an agent,[75] it is he rather than his opponents who is the revisionist.

Not all of Phillips's interpretations of Wittgenstein command universal assent. Thus Fergus Kerr rejects the related view that Wittgenstein intended talk of 'language-games' or of 'forms of life' to apply to religions.[76] However, in the matter of God's existence, Kerr's and Phillips's interpretations run along parallel lines.[77] But significantly Kerr also questions Wittgenstein's grasp of what is at stake when the existence of God is under philosophical or theological discussion, contesting his view that what is at issue is grounds for belief in 'another being like myself, only infinitely more powerful'.[78]

In any case, the difference between belief in God and belief in facts is to be explained, according to Phillips, by the role of belief in God in a person's life, for in the absence of different behaviour there would be reason to doubt that such a belief exists. Belief in God characteristically involves commitment; it is a picture, of which the 'use' or 'method of projection' must not be ignored. And the

---

[73] D.Z. Phillips, 'On Really Believing', in Runzo, *Is God Real?*, 85–107, pp. 85–6, 97.

[74] Phillips (1970), *Faith and Philosophical Enquiry*, p. 85; see also pp. 130–31, and Phillips (1965), *The Concept of Prayer*, Chapters 1 and 2.

[75] Phillips, *Faith and Philosophical Enquiry*, pp. 130–32. For the centrality in Christian theology of the belief that God is an individual, see R.W. Hepburn (1963), 'From World to God', *Mind*, **72**, 40–50, p. 42.

[76] Fergus Kerr (1986), *Theology after Wittgenstein*, Chapter 2.

[77] Kerr, ibid., pp. 151–6.

[78] Kerr. ibid., pp. 153–4.

'method of projection', for Phillips and for the Wittgenstein of his interpretation, is such that the language of existence is inapplicable to God. Rather, 'All things have their meaning in God, for the believer. That is why God is not a maker, but a creator.'[79] Here Phillips reaffirms that belief in God's eternal being 'before ... even thou hadst formed the earth and the world' cannot be understood 'independently of the method of projection in which' such beliefs 'have their natural home'.[80] Thus the language of creation, once its projected nature is grasped, concerns not causal dependence but meaning.

Commenting on Phillips's essay, John Hick accepts that 'To believe in God and yet live as though there is no God is not genuinely to believe in God', but is puzzled that Phillips thinks that the realist would say (or has to say) otherwise. The realist can endorse a dispositional account of belief as readily as the non-realist.[81] But as Hick goes on to say, the question of what it is to have a certain belief needs to be distinguished from that of what it is for that belief to be true. Where Phillips seems to hold that belief in God lacks a referent and that its truth consists in its being believed, the realist holds that 'the existence or non-existence of God is a fact independent of whether you or I or anyone else believes that God exists. If God exists, God is ... the ultimate creative power of the Universe'.[82] For, as Hick writes in reply to another symposiast, 'the theistic belief ... is belief in a God who exists independently of human belief and whose goodness and ultimate sovereignty give final meaning to our lives'.[83] I have heard it suggested that all theistic religious language is susceptible to an anti-realist interpretation; some of the remarks just quoted would seem soundly to confute this view. Insofar as anti-realist positions such as that of Phillips reject the conviction that God exists independently of human belief and is the ultimate creative power of the universe, they seem just as revisionist as positions like that of Braithwaite that Phillips castigates on this count.[84] But unless such convictions are incoherent, they should be allowed consideration by philosophers of religion, in case grounds can be found for accepting them.

However, Phillips has subsequently given support (albeit with qualifications) to claims of Ludwig Feuerbach that such convictions are indeed incoherent. In *The Essence of Christianity* (1841), Feuerbach is clearly concerned with issues of coherence, since he argues that attributes ascribed to God such as love and understanding, which are properly human attributes, make no sense when their

---

79 Phillips, 'On Really Believing', p. 105. Phillips seems to disregard over a score of biblical references to God as 'maker', for example of heaven and earth.

80 Phillips, ibid., p. 105.

81 John Hick (1993), 'Believing – And Having True Beliefs', in Runzo, *Is God Real?*, 115–16, p. 115.

82 Hick, ibid., p. 105.

83 Hick (1993), 'Belief in God; Metaphysics and Values', in Runzo, *Is God Real?*, 130–32, p. 132.

84 Phillips, 'On Really Believing', p. 97.

human connotations are removed. But, as Byrne replies, this is to assume that there is no alternative besides the use of these attributes in either an anthropomorphic or an equivocal sense,[85] a view that I have already contested in Chapter One. Feuerbach also equated real subjects with empirical subjects, whose attributes are contingent and who might have lacked them, and here Phillips endorses his reasoning and his conclusion that God is not a metaphysical subject (conceived on the model of such empirical subjects).[86] However, as Byrne remarks, God might be like empirical objects in being a subject of predication, but unlike them in other respects. (Byrne here suggests that creation is essential to God and that God could not forego the property of being creator and continue to be God,[87] but this suggestion seems to abandon the traditional belief that God could exist even if there were no world, and to remove from God the freedom to create or not to create.)

In support of Feuerbach, Phillips now produces an argument against the coherence of God transcending the world. The universe cannot intelligibly be regarded as a thing, of which God might be the source, for there is no basis on which to individuate such a thing; nor can it be regarded as the class of things, for there can be no such class as the class of things that exist (everything), and no criterion of identity for such a class.[88] Byrne proposes in reply that 'What is meant by the world when it is said that God transcends the world is the cosmos ... the collection of spatio-temporal objects thought of as forming an ordered physical whole'.[89] While this is a promising reply, anti-realists might reject the claim that these objects comprise an ordered whole,[90] claiming, as Kant did, that the order could be read into the world by ourselves.

However, the world could instead be construed as simply the set of spatio-temporal objects, as William Rowe regards it in preparing the ground for a version of the cosmological argument.[91] It would then be open to us, as Byrne suggests, to ask whether this set having the members it does, and indeed having members at all, is self-explanatory or stands in need of an explanation beyond itself. We would not then be seeking an explanation for everything by reference to something else somehow situated over and above everything (Phillips's objection to the

---

85 Ludwig Feuerbach [1841] (1957), *The Essence of Christianity*, trans. M.A. Evans (George Eliot); Byrne, op. cit., p. 130.

86 Phillips (2001), *Religion and the Hermeneutics of Contemplation*, pp. 94–5.

87 Byrne, op. cit., p. 132.

88 Phillips, *Religion and the Hermeneutics of Contemplation*, p. 70. See also *Faith and Philosophical Enquiry*, p. 42.

89 Byrne, op cit., p. 132.

90 See Cupitt, *Creation Out of Nothing*, p. 53.

91 William L. Rowe (1970), 'Two Criticisms of the Cosmological Argument', *The Monist*, **54**, 441–59.

cosmological argument),[92] but rather an explanation of the set of material or physical objects, and (perhaps) of the set of the states of such objects, by reference to something immaterial with the power to bring them into being or bring them about. Since these remain intelligible enquiries, the objections of Feuerbach and Phillips to realist understandings of belief in creation are inconclusive. But this is not the place further to debate issues surrounding creation, which are addressed in Chapter Four.

Since the arguments for global anti-realism and against global realism are inconclusive, and so are the arguments for religious anti-realism and against realism in theistic varieties of religion (such as the arguments of Phillips), there is every reason to adhere to such a realist understanding in philosophy of religion as well as in other sectors of philosophy. Realism is the obvious interpretation of theistic religious language (as most atheists and agnostics would agree), and a strong argument would be required to justify its abandonment. Such realism should be of a critical kind, as advocated by Arthur Peacocke, since claims to religious truth need to be sifted, and far from all are even coherent;[93] but epistemological caution (the 'critical' component of 'critical realism') is compatible with the belief that religious language is intended to refer to a reality independent of us and of our conceptions (the 'realism' component). Adherence to realism is all the more important since, as has emerged above, some of the claims made by believers in creation are ones for which anti-realism has no satisfactory account to offer. These claims are relevant not only to God as a creative agency but also to the possibility of answerability to God, a possibility which, as will be seen in Chapters Nine and Ten, is highly relevant to belief in the human stewardship of nature. We have also encountered ways in which an argument for belief in creation could avoid incoherence and might take flight. But first a further challenge to theistic religious language needs to be considered.

---

92 Phillips, *Religion and the Hermeneutics of Contemplation*, p. 69; Byrne, op. cit., p. 133.

93 Arthur Peacocke (1990), *Theology for a Scientific Age*, pp. 12–19.

Chapter Three

God and Falsification

Section 1: Popper, Duhem and Quine

In this chapter, falsification is brought to bear on theistic belief, and shown to throw light on its logic. The method of falsification was first applied to religious language by Antony Flew in the 1950s, but the implications of falsification had already been discussed among philosophers for several decades with relation to the philosophy of science. This background needs first to be introduced, so that the scope and the limits of falsification can be brought out.

It was Karl Popper who explicitly introduced falsification into philosophy in 1919-20. Wishing to distinguish between valuable theories such as Einstein's theory of gravity and what he considered less fruitful ones such as psycho-analysis and the Marxist interpretation of history, he formed the view that what gives a theory merit is its refutability. 'A theory which is not refutable by any conceivable event is unscientific. Irrefutability is not a virtue in a theory (as people often think) but a vice.'[1] Whenever a theory is reinterpreted so as to rescue it from refutation, as Popper found adherents of Marxism and of psycho-analysis prone to do when apparent predictions proved false, the cost is the destruction of its scientific status. Popper summarized his stance as follows: 'the criterion of the scientific status of a theory is its falsifiability, or refutability, or testability'.[2]

In Popper's work, the theory of falsifiability was never a theory of meaning. A decade later he represented it as a criterion of demarcation: 'statements, or systems of statements, in order to be ranked as scientific, must be capable of conflicting with possible, or conceivable, observations'.[3] Popper did not, however, regard all non-scientific statements as worthless. For some myths are capable of being developed and becoming testable, and most scientific theories originate from such myths. Examples of myths containing anticipations of scientific theories include Empedocles' theory of evolution by (natural) trial and error, and Parmenides' myth of an unchanging universe, in which Popper found an anticipation of Einstein.[4] Popper expressed his ideas in an unpublished work of 1931–1932, which much later became his *Logic of Scientific Discovery* (1959).[5]

---

1 Karl R. Popper (1963), *Conjectures and Refutations*, p. 36.
2 Popper, ibid., p. 37.
3 Popper, ibid., p. 39.
4 Popper, ibid., p. 38.
5 Popper (1959), *Logic of Scientific Discovery*.

He also presented them to the Vienna Circle, explaining the related limits of verification, but stood aside as the participants attempted to modify their theories to cope with such problems.[6]

Falsification has problems of its own. Thus while universal generalizations can be falsified, existential affirmations such as 'there are leprechauns' cannot. Besides, as Ayer remarked, despite the difficulties inherent in inductive methods, scientific reasoning is sometimes concerned with confirmation as well as with disconfirmation.[7] However, a more serious problem arises because, at least in practice, single hypotheses or theories cannot be falsified one by one. Belief in the flatness of the Earth is not conclusively falsified by the way ships appear to sink below the horizon, since light might not travel in straight lines, or there could be troughs in the sea, or sea air could adversely affect eye-sight. What can be falsified are clusters of beliefs, often including statements about initial conditions and the powers of observers, in addition to the hypothesis in question. This had already been pointed out by Pierre Duhem in 1914.

Duhem argued that an experiment cannot test a single hypothesis alone, but rather an entire theoretical scaffolding. Thus when an experiment disappoints our expectations, 'the only thing the experiment teaches us is that among the propositions used to predict the phenomenon … there is at least one error; but where the error lies is just what it does not tell us'.[8] Thus it is whole bodies of theory that have empirical content (holism), and this is a problem both for verificationists and, more directly, for adherents of Popperian falsification. Duhem's insight was later taken up by Quine, who remarked in 1953 that 'Our statements about the external world face the tribunal of experience not individually but only as a corporate body'.[9] While Quine was raising a problem for his fellow-verificationists, the Quine/Duhem thesis also requires falsificationism to be significantly modified, so that what is falsifiable will often be not a single statement but the relevant cluster.

Quinean holism, however, suggests that any of our beliefs could be fallible, including ones previously considered analytic. Hence Quine proceeded to represent the epistemology of confirmation and disconfirmation along the lines depicted by Misak: 'Our belief system is like a web, with mathematics and logic at the centre, gradually shading into highly theoretical hypotheses and then into sentences about particular observations at the periphery.'[10] Any hypothesis can be confirmed or disconfirmed, depending on where, in the face of problems deriving from experience, we choose to make adjustments. (We can set on one side here the issue of whether this is really a matter of choice.) However, in reply to Quine's

---

6 Popper, *Conjectures and Refutations*, p. 41.

7 A.J. Ayer, *The Central Questions of Philosophy*, pp. 27–30.

8 Pierre Duhem [1914] (1954), *The Aim and Structure of Physical Theory*, p. 185.

9 W.V.O. Quine (1953), 'Two Dogmas of Empiricism', in *From a Logical Point of View*, p. 41. See further Sandra G. Harding (ed.) (1976), *Can Theories Be Refuted?*

10 C.J. Misak (1995), *Verificationism: Its History and Prospects*, p. 145.

adoption of a holistic view of meaning and implicit suggestion that the meaning of no sentence can be known in isolation from the rest of a speaker's beliefs, Dummett replied that this stance undermines all theories of communication,[11] later adding that, despite the Quine/Duhem thesis, the assignment of truth-values to particular sentences can still be mandatory.[12]

Parallel responses would clearly be open to falsificationists, who could give predictions of the weather as examples of predictions amenable to determinate truth-values in light of experience. Thus is it possible to grant much to the Quine/Duhem thesis (holism as introduced above), without endorsing the sceptical implications about meaning derived by Quine from that thesis. By the same token, while that thesis requires falsificationism to be modified in the manner just specified, and may also suggest that falsification need not be of a sensory character,[13] the abandonment of falsificationism is not in question.

Section 2: The Falsification Controversy

Antony Flew's article 'Theology and Falsification' introduced falsification as a condition of demarcation for statements. Flew's opening parable, borrowed from John Wisdom, shows how 'a fine brash hypothesis' can cease to amount to an assertion through an unchecked process of adjustment after adjustment, dying 'the death of a thousand qualifications'.[14]

Two explorers find a clearing in a jungle, with 'many flowers and many weeds'. One maintains that it is tended by a gardener, while the other denies this; and so they pitch their tents and watch, without event. In case the gardener is invisible, they set up a barbed-wire fence, electrify it, and patrol with bloodhounds, but still without the least evidence of a gardener emerging. Yet the Believer persists in holding that there is a gardener who is invisible, intangible, insensible to electric shocks, who has no scent and makes no sound, but comes secretly to tend the garden. To this, the Sceptic eventually asks how such a gardener differs from a gardener who is purely imaginary.[15] Flew recognizes that such a process of qualification can be halted in time for an assertion to remain (if, say, the claim were that the gardener is invisible but tangible). But would-be assertions like those of his Believer have ceased to be assertions or statements at all.[16]

To such would-be assertions, Flew likens theological utterances such as 'God has a plan', 'God created the world' and 'God loves us as a father loves

---

[11] Michael Dummett (1973), *Frege: Philosophy of Language*, pp. 598–9.

[12] Dummett, 'Reply to Brian Loar', in B. Taylor (ed.), *Michael Dummett: Contributions to Philosophy*, 269–80, p. 273.

[13] Misak, p. 152.

[14] Antony Flew (1955), 'Theology and Falsification, A', in Antony Flew and Alasdair MacIntyre (eds), *New Essays in Philosophical Theology*, 96–9, p. 97.

[15] Flew, ibid., p. 96.

[16] Flew, ibid., p. 97.

his children'. They look like 'vast cosmological assertions', but they seem to deny nothing. Yet to assert something is to deny its negation. Indeed whatever would count against the assertion comprises at least part of the meaning of that negation, and knowing the meaning of the negation amounts in effect to knowing the meaning of the assertion. Thus if nothing is denied, then nothing is asserted either; and this is just what the Sceptic in the parable suggests about the Believer's statement.[17] Flew's challenge to believers thus involved asking what conceivable state of affairs would have to happen to constitute disproof of the love, or the existence, of God.

We can set on one side the small ingredient of verificationism and of semantic theory here, present when Flew suggested that knowing what would count against an assertion amounts to knowing the meaning of its negation. His key point is that where nothing is denied, nothing is asserted either, and it is an entirely reasonable one. What is denied need not always be something empirical, and here the parable of the gardener may be misleading;[18] but the request for believers to specify what is denied, and thus what would count as the falsification of their assertion, remains appropriate.

This application of the issue of falsifiability as a criterion of statementhood, rather than of statements having a scientific status, was an innovation on Flew's part; for Popper did not take the view that claims of psycho-analysis or Marxist were not statements, nor that no statements were present in the pre-scientific mythical claims of Empedocles or Parmenides. However, Popper could agree with Flew that something would have to count as their claims being false if those claims were to amount to statements of any kind. For example, for Empedocles' claims about evolution to amount to statements, they would have to exclude some possible state of affairs such as life-forms evolving without any process of trial and error. This state of affairs would not need to be shown to be actual, but unless it could be described without self-contradiction as a possibility, there would be no statement present in Empedocles' claim. Popper could indeed be said to rely on such possibilities, in holding that intellectual progress is made when attempts are made to subject claims or conjectures to tests in which they might possibly be falsified. This is what he construes the Greek atomists as having actually done in response to Parmenides' theory of an unchanging universe.[19]

On the other hand, not all states of affairs that would falsify statements would need to be empirical, nor contingent. Thus for Goldbach's Conjecture to be falsified, an even number greater than 2 would have to exist that was not the sum of two primes, but the existence of such a number could be neither an empirical nor a contingent state of affairs. And even if mathematical claims are set aside, most historical claims can no longer be falsified in the present, even though there are historical contingencies of the past that they exclude. So Flew's

---

17 Flew, ibid., pp. 97–8.

18 Dan R. Stiver (1996), *The Philosophy of Religious Language*, p. 49.

19 Popper, *Conjectures and Refutations*, pp. 81–3.

test of statementhood has to be interpreted accordingly, and cannot reasonably require empirical falsifiability for statements of every kind. Further, acceptance of the Quine/Duhem thesis suggests that claims cannot invariably be considered for falsification one by one, rather than in clusters, and this too could be relevant to claims of a religious kind.

One of the responses to Flew, that of R.M. Hare, has been mentioned already in Chapter Two. Hare granted that Christian talk of God does not involve statements, and suggested instead that it involves expressions of fundamental commitment.[20] To this anti-realist account, Flew replied, in a robust realist mode, that such an interpretation fails to capture what believers intend, and also fails to provide for the job that such religious language is supposed to do, which includes supplying grounds for other beliefs and reasons for religious practices.[21]

The task of defending realist interpretations and claims about God involving statements was taken up rather by Basil Mitchell. Mitchell considers that Flew's suggestion that theological claims have become too qualified to be statements because they rule out nothing implies that the problem of evil is not a problem for the theist. But the problem of evil really is a problem.[22] (Significantly, Cupitt now represents it as no longer a problem for his anti-realist religion;[23] if God is not the creator of the material universe, then the problem disappears as a distinctive problem for religious believers.) Suffering, as Mitchell says, counts against belief in God, but the believer does not regard it as counting decisively, unless it is irreconcilable with God's will and goodness.[24]

This is the setting of Mitchell's parable of the Stranger. A Partisan of the resistance in an occupied country in time of war meets one night a Stranger who deeply impresses him, explains that he commands the resistance, and urges the Partisan to have faith in him whatever happens. The Partisan is convinced. Subsequently the Stranger is sometimes seen helping members of the resistance, and sometimes handing over patriots to the occupiers. When the Partisan asks for help, sometimes he receives it, and sometimes not, but his trust is undimmed. But his friends sometimes ask what the Stranger would have to do for the Partisan to discard his faith and admit that the Stranger was not on their side.[25]

Mitchell's parable is more directly relevant to 'God loves human beings' (particularly when held in a Christian or Jewish context) than to 'God created the world'. The adverse evidence could be held not to count decisively against God's love because (although Mitchell does not spell this out) a loving God might favour human beings having the kind of freedom that makes possible deeds with

---

20 R.M. Hare, 'Theology and Falsification, B', in Flew and MacIntyre, 99–103.

21 Flew, 'Theology and Falsification, D', in Flew and MacIntyre, 106–8, p. 108.

22 Basil Mitchell, 'Theology and Falsification, C', in Flew and MacIntyre, 103–5, p. 103.

23 Don Cupitt, *Creation Out of Nothing*, pp. 60–67.

24 Mitchell, pp. 104–5.

25 Mitchell, pp. 103–4.

harmful consequences, and for the sake of this same freedom might refrain from preventing such events. As for evils not caused by human action or forbearance, these could be held to stem from natural regularities of the kind that a loving God would also favour in order to allow meaningful action on the part of creatures to be possible. Hence God's love may not conflict with the occurrence of the evils that actually happen, whether they have a human origin or not. (These topics are discussed in greater detail in Chapter Seven.) What would conflict with God's love would perhaps be pointless evils, but whether there are such evils involves an appraisal, as this reasoning indicates, both of actual events and of clusters of beliefs about God. Since something could conflict with the claim about God's love, it remains a statement.

Further, the parable of the Stranger brings in the evidence of the past impression made on the Partisan, and this could be held to correspond to evidence deriving from the life of Jesus Christ, and/or from the Exodus and the lives of the prophets; only if new findings were to overthrow such evidence would it cease to be reasonable to retain a related trust. This issue brings in the possible falsification of claims based on history; but until the evidence of the past is undermined (something yet to happen), faith analogous to that of the Partisan retains at least some of its grounds. Mitchell has substantially endorsed this expansion of his thinking.[26]

Responding to Mitchell, Flew agrees with Mitchell that the problem of evil counts against the assertion that God loves people, but contends that the attributes ascribed to God by believers rule out explanations that reconcile his love with the facts of experience; the theologian may well then be driven to qualify the assertion about God's love, instead of explaining it.[27] Issues of theodicy need to be reserved for discussion of natural selection (see Chapters Six and Seven below), but Flew's admission amounts to a recognition that there are theological statements, and that these are fit for appraisal on a realist basis. And this recognition supplies the present section with its conclusion.

It may be worth while to pursue separately the issues of what could falsify belief in creation, and also of what can be learned from the falsification debate about statements. These issues will now be discussed in turn.

Section 3: Creation and Falsification

What could falsify belief in creation? One of the religious claims mentioned by Flew is expressed as 'God created the world'. But creation cannot be regarded as a past event, and concerns, as we have seen, the dependence on God of all material

---

26 Mitchell, private communication, 13 October 1971. Mitchell there endorsed, though with slight qualifications, a version of the interpretation given here, as supplied in Robin Attfield (1970), 'Non-Tentative Religious Beliefs and Rationality', *Sophia*, **9**, 16–21.

27 Flew, 'Theology and Falsification, D', p. 107.

objects at all times. Accordingly the claim can be better expressed as 'God is the creator of the world'. Here 'the world' refers not to everything but to the set of material objects and of their states.

For God's agency to be capable of bringing it about that these objects exist rather than others and rather than none at all, God has to be unlike them in being unable to come into existence or to cease to exist; and for God's agency to be capable of bringing it about that they have the states that they do and not others, God has to be unlike creatures in not being susceptible to change initiated externally. (Since the explanation of those states of creatures which result from the free actions of creatures is more complex than the explanation of other states, it is simpler to exclude these states from consideration in this context; however, they are not altogether irrelevant, since a creator would in any case have the power to change them.) God would also have the power and knowledge necessary to bring into being whatever can be brought into being without contradiction, with obvious exceptions such as things so described that they are expressly beyond God's control. Indeed the term 'God' is so used here that to be God is to have these capacities and incapacities. Relatedly talk of God's wisdom (or, as in Flew's other example, of God having a plan) should be understood as concerning a creator of this kind. Such talk does not make God 'a being among other beings'; for to have such a nature as this is to transcend all other beings not only in quality or through existing independently of them but also in being able to bring them into being.

How, then, could the claim that God is the creator of the world (understood as above) be falsified? This question seems unanswerable, for, in the nature of the case, nothing that could happen to things in the world would count as evidence against it. Nothing less is to be expected; for the claim that God is the creator of the world is introduced, in part, to explain there being material things rather than there being none. Further, if this claim really does explain there being material things, then, rather than being an empty assertion it is all-important; for (so claim its adherents) if it had not been the case, then there would have been no material world.

This being so, further progress may be possible with regard to falsification. For, as we have seen, falsification is sometimes relevant to clusters of claims; and the claim about creation could be paired with a claim about explanations, along the lines of what has been called since the time of Leibniz 'the Principle of Sufficient Reason'. Many possible versions of such a principle have been considered; one of the more cogent was presented by Robert Nozick, and amounts to the claim that there is a sufficient reason for whatever could have been otherwise, except where there is a sufficient reason to the contrary.[28] While no such principle is likely to be recognized as true *a priori* (or on any other basis), and therefore no argument

[28] Robert Nozick (1981), *Philosophical Explanations*, pp. 140–42, 671–4. Thus qualified, the Principle is immune to the objection to unqualified versions from Antony Flew (in Flew (1966) *God and Philosophy*, at pp. 83–4) that since the ultimates of every system are invariably inexplicable, the Principle is demonstrably false.

based on such a principle is going to be of unquestionable soundness, there is a strong case for holding that ordinary reflection presupposes just such a principle as Nozick's, and that this view is relevant not only to the causation of events but also to the bringing into being of entities that would not exist otherwise.[29] In any case, what is required for attempts at falsification to begin need not be secure knowledge; a cogent claim governing explanations suffices, and can be found in the principle just put forward.

When this form of the Principle of Sufficient Reason is conjoined with the fact that the set of material objects to be found around us is actual (or has members), the implication is that this state of affairs has an explanation; and while some explanations would, by the same token, require ulterior explanations, this is not the case with the possible explanation in terms of a creator who or which cannot be brought into being or destroyed or changed by anything exterior. Thus, given the Principle of Sufficient Reason as understood here, there being material objects (in the same sense) must have an explanation, but the existence of such a creator is not subject to this requirement. Hence the existence and agency of such a creator explains the existence of the set of material objects, without requiring a further explanation. This reasoning is, of course, a version of the cosmological argument (considered further in Chapter Five, Section 2). No suggestion is made that this argument is conclusive. But its mention in the current context should not evoke surprise; for it has been maintained that all those who attempt to move from the material world of finite objects to a God who is not identical with that world but transcends it are committed to an argument of this general kind, even if they consciously reject it.[30]

If we now return to the issue of falsification, and ask what follows if the claim that God is the creator of the material world is false while the Principle of Sufficient Reason as here understood is granted, we find that the implication is that there is no material world. For if there were such a world, there would be an unexplained state of affairs subject to a principle which denies that there are such unexplained states of affairs, and this is impossible. (This reasoning figures again in Chapter Five, Section 5.) This finding could be expressed by saying that, given commonplace assumptions about explanation, the claim about creation makes a world of difference.

It also shows that even the claim that God is the creator of the world can be regarded as open to falsification, and thus as comprising a statement.[31] The falsification could not, in the circumstances, be empirical, since if there were no creator, and given the same assumptions about explanation, there would (for the same reasons) be no sentient creatures, let alone any rational ones, to have

---

[29] William L. Rowe (1971), 'The Cosmological Argument', *Noûs*, **5**, 49–61, pp. 59–61.

[30] R.W. Hepburn, 'From World to God', p. 43.

[31] For a parallel argument for the falsifiability of theism, see Keith Ward (1996), *God, Chance and Necessity*, p. 98.

experiences of any sort. But we have already seen that not all falsification can be of an empirical character. The possibility that there would be no material world is in any case sufficient to show that, given principles such as the Principle of Sufficient Reason (or indeed given the mere cogency of such principles), the claim about creation can involve a significant assertion; for that claim could have been false, and (given the same Principle) there would then have been nothing. And this tells us a good deal about the logic of belief in creation, as well as about its falsifiability.

Section 4: Falsification, Religion and Explanations

What can be learned from the falsification debate about religious statements and about explanations in general? The falsifiability in theory of claims about creation, even though these claims transcend sensory experience, opens up the possibility that many other theistic claims are also falsifiable and thus satisfy the requirements for statements. Some such claims, however, are falsifiable in another way. For claims concerning God's concern for humanity are often taken to imply that God would confer revelations to human beings at some stage or other, and, if so, would be falsified if there are no revelations. So if there are revelations, whether through Mohammed or Christ or Moses, these claims pass a relevant test. Admittedly it would be difficult to establish that there are no revelations; but there could certainly be evidence both for and against this view.

Purported revelations are beset with problems, such as that of the supposed revelation being expressed in language which may be ambiguous or embody misunderstandings, or being conveyed through imperfect processes of transmission or translation. Even if the original recipient or recipients grasp an unblemished message, it may reach others in an imperfect condition. It may then require all the resources of rational interpretation for a message of any value to be discerned. But for present purposes these problems are beside the point. Even if nothing better than a fragmentary revelation reaches humanity in general, that would be sufficient to confute the suggestion that there are no revelations. Nor is there any need to argue here that there actually are revelations of even a fragmentary character. For the possibility that there are suffices to show that many theistic claims (such as ones expressing divine concern for humanity) embody statements. As such, they could even be included somewhere in Quine's web of beliefs capable of confirmation or disconfirmation (see Section 1 above).

Theistic claims could also in principle be vindicated by experiences of irresistible theistic disclosures happening after a person's death, although in this case no evidence would be available to the living of such experiences either happening or not happening. John Hick presents, in response to Flew, a scenario in which such an experience could happen without apparently any contradiction being involved in its description.[32] Hick also stresses that such encounters would

[32] John Hick, 'Theology and Verification', pp. 58–69.

involve verification rather than falsification, but cogently maintains that their occurrence would still count in favour of relevant theistic beliefs.[33] But this possibility has relevance to falsification too, for those who claim that God confers life and illumination after death would probably agree that this claim excludes the possibility that there are no such experiences. Because a possible state of affairs is excluded, this of itself allows such a claim to count as a statement, albeit one for which the relevant evidence is unavailable during a person's natural life.

In the light of stances like that of Hick on the limits of falsification, it is worth returning to more standard circumstances and asking whether falsification can continue to play the role for which Popper cast it, without casting doubt on Flew's claim that a statement necessarily involves the denial of its negation, and must for that reason be falsifiable, at least in principle. In particular, it can be asked whether explanations need to be empirically verifiable or falsifiable.

In this connection, Ayer eventually took the view that, in addition to primary systems containing propositions which describe what can actually be observed, there can be secondary systems which are meaningful provided they have 'some explanatory value'.[34] The idea is, in part, that such claims, when combined with accepted facts, allow of empirical predictions that are not available without them. The philosophy of science certainly needs to provide for explanations of which this is true.

Whether Popperian falsification makes sufficient provision for such explanations is unclear. It provides for those that could be falsified, but seems not to provide for ones admitting of empirical confirmation but not of empirical disconfirmation (like 'there are black holes'). It can afford to grant that falsification can be of clusters of claims, such as an explanation plus statements of initial conditions and of observers' powers. But in supplying a demarcation criterion for science it cannot afford to discard the requirement that falsification be empirical, the requirement which was rejected above with regard to falsification as a test of statementhood (whether in science, history, mathematics or other fields). Otherwise it could provide for hypotheses that can be confirmed but not disconfirmed; but this gain would admit too much to the scope of science, and would thus be purchased at too heavy a cost. What may be needed is the inclusion of the possibility of empirical confirmation as well as that of empirical disconfirmation as another mark of a scientific hypothesis, plus constraints to avoid this admitting (for example) astrological hypotheses.

The situation has parallels when the issue is that of the requirements for statements in general. Here we can no longer require empirical confirmation or disconfirmation. But we can reasonably require that those making claims have either reasons in favour of what is claimed, or reasons against its negation. For to assert a claim is to give it one's backing, and this is only intelligible if the speaker or writer concerned has reasons of one of these kinds, whether they be

33 Hick, ibid., pp. 57–8.

34 Ayer, *The Central Questions of Philosophy*, p. 33; Misak, p. 77.

rules, principles, definitions, facts, or experiences. Misak elicits a similar account from the views of Moritz Schlick; verifiability is to consist in a speaker or writer having reasons in favour of a hypothesis.[35] While this is not necessary for meaningfulness, it seems sufficient (although not necessary) for statementhood. To supply a necessary condition, a disjunction is needed of verifiability (in this sense) and of falsifiability (in the matching sense of having reasons against the negation of one's claim).

This view includes more than an echo of Flew's stance about when a claim is intelligible as a statement or assertion. It is also a view that he would be free to adopt if challenged about his recent endorsement of belief in a deity, albeit an Aristotelian or deistic one, and not one conceived along traditionally theistic lines. The existence of such a deity is held by Flew to help explain the origins of life.[36] It is not completely clear that this hypothesis facilitates predictions that are unavailable in its absence, but if the probability of life arising is held to be greater on this hypothesis than in its absence, that would count as a reason in its favour and allow it to be regarded as a statement on the basis just introduced.

It emerges that there are multiple ways in which religious claims prove falsifiable in theory, thus satisfying the falsifiability requirement for comprising statements. If explanations (or systemic clusters of them) are to count as scientific they must apparently be (at least) open either to empirical confirmation or disconfirmation; but for statements in general (religious statements included), no such empirical requirement is in place. Rather, those making claims must either have reasons in favour of what is claimed, or reasons against its negation, for their claims to amount to statements. On this basis, the class of statements turns out to include Flew's recent claims about God.

Section 5: Findings So Far about Creation

While the main stress of the first three chapters has been on the meaningfulness of theistic language, claims and statements, and on theories of meaning and understanding in general, nevertheless creation has been another recurrent theme. Belief in creation may be held to transcend verification, but creation is also related to experience in that it is believed to be what makes experience possible, and therewith the phenomena that science studies (Chapter One, Section 2). In the nature of the case, it lies beyond experience. Yet if it explains the possibility of experience, its lying beyond experience does not count against it. Further, as has been seen in Section 3 above, this is a belief which, granted cogent (if unproven) principles of explanation like the Principle of Sufficient Reason, could in principle be falsified, not through one experience rather than another, but on the simple basis

---

35 Misak, p. 81; cf. Moritz Schlick (1976), 'Meaning and Verification', in Schlick, *Philosophical Papers, Volume II: 1925–1936*, p. 480.

36 Flew, 'On Darwinism and Theology', and Flew, 'My "Conversion"'.

that there being a material world can be confirmed or in principle disconfirmed against the test of experience. (A realist account of the material world would supply the most natural interpretation of the argument just presented, albeit not the only possible one; relevantly, all the main arguments for generic anti-realism, including anti-realism about the material world, have been considered in Chapter Two and found wanting, and that makes the adoption of such an account, already the most natural interpretation, all the more obviously legitimate.)

Given the strength of the objections to verificationism, that theory of meaning cannot be considered to undermine belief in creation (for it misconceives the relations of meaning and verification, and contrives to undermine itself: see Chapter One, Section 5 above), even if it were not the case that weak verificationism seems to provide for the possibility of experiences which could count indirectly in favour of this belief (Chapter One, Section 4). The real problems about the meaning of theistic language relate rather to that of avoiding the twin dangers of using language anthropomorphically and equivocally; but a solution to this problem has been argued to be supplied in the theory of analogy (Chapter One, Section 6). If language about the dependence of the material world on an independent source as its creator makes sense, then, as is also argued there, many other claims of theistic discourse can be interpreted as meaningful in this light, and those who use it need not fail to communicate, whether with others or with each other, as some sceptics have suggested.

There could apparently be semantic problems for theistic language if, as generic anti-realists argue, we cannot understand sentences or statements whose truth-conditions transcend verification, and terms such as 'true' have to be understood as having a suitably constrained sense and scope. While some semantic anti-realists such as Dummett do not apply such conclusions to belief in God and creation, others such as Rorty (who considers that realism and belief in creation hang or fall together) do, and it is on the basis of Rorty's semantic and epistemological theories that Cupitt grounds his religious anti-realism. It is thus even more significant that the arguments for generic anti-realism, whether of Dummett, of Wright, or of Rorty, prove inconclusive (Chapter Two, Sections 2 to 4). Hence Cupitt's affirmations that God, the world and the self can only be understood through human conceptions does not show that they cannot be understood as existing independently of those conceptions, or that language about them has to be understood in a highly qualified, anti-realist manner, as if they comprised nothing but an expression of human self-creation (Chapter Two, Section 5).

Phillips's religious anti-realism has a distinct basis, concerning religious language rather than general semantics, but his anti-realist arguments also seem inconclusive (Chapter Two, Section 6). His account of creation as what gives meaning to the believer and the believer's world seems not only not to preclude a realist understanding of that language, but to become much more cogent if such an understanding is adopted; for an awareness of dependence on God's creative activity could then underpin a recognition of life as a gift from God (see Chapter Ten, Section 4 below), together with the sense of creatureliness mentioned in

Chapter One, whereas an anti-realist approach suggests that there is no creative agent to whom such gratitude is possible. (This becomes relevant again when the issue of transcendence is discussed in Chapter Four.)

Popper's falsification test was intended as a criterion demarcating scientific claims (Chapter Three, Section 1), and thus of no direct relevance to belief in creation (except insofar as this belief seems historically to have stimulated scientific research; see Chapter One, Section 6 above). Indeed Popper finds room for many conjectures and hypotheses (those of Empedocles and Parmenides included) that he would have regarded as empirically unfalsifiable. However, Flew's adaptation of falsifiability as criterial for statements (and for sentences intelligible for use as statements) was importantly correct, but he was mistaken in requiring falsification to be empirical, and (like Popper) in expecting claims to be falsifiable singly rather than in clusters. Thus it turns out (in the light of reflection on Flew's and Mitchell's parables) that theistic claims such as 'God loves us' are falsifiable, and hence comprise statements (Chapter Three, Section 2); as, on reflection, claims about creation must also be acknowledged to be (Chapter Three, Section 3). This could also apply to claims about revelation, and to the kind of non-theistic belief in God recently adopted by Flew himself; for those making claims must simply have either reasons in favour of what is claimed, or reasons against its negation, for their claims to comprise statements (Chapter Three, Section 4). Further, if we endorse principles of explanation such as the Principle of Sufficient Reason, then claims about creation are just as falsifiable as the claim that there is a material world; and if we reason from the latter to the former, adopting a realist interpretation of belief in creation, we find ourselves involved in cosmological reasoning, which has been suggested to be implicit in all moves from the world to a creator considered as independent of it (Chapter Three, Section 3).

But we have yet to investigate how belief in creation diverges from alternative theologies, how it relates to the traditional theistic arguments, and how it differs from the stance with which it is frequently confused, creationism. These (together with theories of evolution and their relation to belief in creation) are matters considered in the second part of this book, that on Creation and Evolution.

PART II
Creation and Evolution

Chapter Four

Creation

Section 1: Some Alternative Conceptions

Belief in creation is far from the only form of belief in God, and creation is far from the only relation possible between God and the material world. As we have seen in Chapter Three (Section 3), for God's agency to be capable of bringing it about that the familiar physical objects around us exist rather than others and rather than none at all, God has to be unlike them in being unable to come into existence or to cease to exist. By the same token, God has to exist independently of physical objects. But there have been many other concepts of God for which God does not create physical objects out of nothing, or is not entirely unlike them, or entirely independent of them. The relevant ones to be mentioned here are Platonism, pantheism and minimalism; brief mention also needs to be made of panentheism.

In Plato's dialogue *Timaeus*, there is a designer (the demiurge), who does not create the material world but imposes order on chaotic matter that exists independently of his activities.[1] This conception supplies an explanation of the world's regularities (including those that later came to be called 'laws of nature'). But it also represents its deity as finite in power, for he has no control over the existence of matter; indeed the order that he imposes is less than perfect because of this lack of control. (Hence in the history of thought, this conception has tended to discourage an empirical approach to the study of nature, as opposed to a study of those mathematical paradigms that it would embody if it were more compliant with the divine will.)[2] It also offers no explanation of the existence of physical objects, as opposed to the order that they display. Besides leaving unanswered the cosmological question of why such objects exist, this stance has widely been considered to make the deity of this conception unfit for worship, being limited in

---

[1] Plato (1929), *Timaeus*, in Plato, *Timaeus, Critias, Cleitophon, Menexenus, Epistles*, trans. R.G. Bury, 30 E. The text here expresses the standard interpretation, definitive of 'platonism' in theological contexts. An unusual, more theistic interpretation is offered in David Conway (2000), *The Rediscovery of Wisdom*, pp. 34–52.

[2] Michael Foster (1934), 'The Christian Doctrine of Creation and the Rise of Modern Natural Science', *Mind*, **43**, 446–68; R. Hooykaas (1972), *Religion and the Rise of Modern Science*, pp. 3–5, 29–31, 35–6.

power and insufficiently different from the objects around us.[3] Indeed this stance raises the issue of what would explain the existence of the designer.

About the deity of theism, by contrast, such explanatory questions do not arise, since his or her power is unlimited, and he or she is unable to come into existence or to cease to exist. (See further Chapter Six, Section 3 below.) Further, theistic belief makes it unnecessary to postulate a distinct designer to explain the order among physical objects. If such an explanation is needed, the activity of the creator can supply it, and if the laws of nature have a legislator, the creator can be held to be that legislator. Theistic belief has the further advantages that its deity is entirely unlike the physical objects around us, has unsurpassable power, creates whatever has value, and is fit for worship if anything is. Indeed in the history of thought theistic beliefs eventually led to the empirical study of nature, as the creator's handiwork, combined with efforts to discover the mathematical thoughts which such a creator would have imposed on his or her creation.[4]

In pantheism, by contrast, God is found in nature, and God and nature are sometimes identified (as in the metaphysics of Spinoza). The widespread awe felt towards natural settings and natural forces is focused on nature as a whole, whether conceived as purposive (as in the pantheism of ancient Stoics) or as an indwelling spirit (as sometimes in Romantic poetry). Thus God and the whole system of nature are fundamentally one and the same entity, and this allows spiritual attitudes towards nature to be understood as worship of God at the same time. However, some pantheists combine this stance with regarding God as a nurturing parent or mother, and observable nature as that which is thus nurtured.

To the extent that pantheists identify God and nature, both must have the same properties, and by the same token both have the same modality (contingency or necessity, that is). If nature could have been otherwise, and might not have existed at all, the same will be true of God (on this conception). Hence the same questions that arise about nature, such as why it is as it is, will arise for the pantheist's God; but the pantheist cannot resort to God in order to answer them, as God has now become the entity to be explained. It is partly to avoid problems of this kind that pantheists often hold that God has some properties that nature does not have.[5] But this is to acknowledge God's otherness or qualitative difference from the material world around us. As soon as God is recognized to transcend the world, it becomes possible to tackle the explanatory issues that strict pantheism cannot tackle; but at the same time the pantheist's identification of God and the world has to be discarded. There could still be a close relation between God and the world, but if God makes it what it is (whether in the manner of Platonism or of theism) then God is importantly beyond the world, contrary to the claims of pantheists.

---

3 H.P. Owen (1971), *Concepts of Deity*, pp. 49–59.

4 Foster, ibid.; Hooykaas, ibid., pp. 1–53; Robin Attfield [1978] (1993), *God and The Secular*, pp. 15–67.

5 Owen, op. cit., p. 75.

The term 'panentheism' has been devised for a range of stances according to which God is present in the world, without the two being identical.[6] Panentheism is often welcomed as stressing God's immanence (or in-dwelling presence) in the world, and thus fostering recognition of its value. In one version of panentheism, God is the creator, but is also regarded as present in, or inhabiting, his or her creation; if this stance can be held consistently, then theism is capable of incorporating some of the features that attract people to pantheism. But in other versions, God is regarded as finite, and subject to change just as natural creatures are. Whatever the advantages of such a position, it has the disadvantages that (for much the same reasons as were introduced above about Platonism) God can no longer be regarded as the creator, or considered as sufficiently different from creatures to be fit for worship. It is thus difficult to generalize about panentheism, as care is needed to discover the particular implications of particular versions. But its possibilities remain important, since the possible criticism that the creator of theism must seem remote from his or her creatures can be countered if theism is compatible (as is argued below in Chapter Eight, Section 5) with belief in the immanence that would attach to the God of panentheism.

A very different, minimalist account of creation is offered by Peter Atkins in *The Creation*. Atkins assumes that the created order is maximally simple, and infers that the creator's job was so slight that he or she 'might as well not have existed'.[7] In this account of creation, there is no role for a creator, and order develops out of nothing without further explanation. But such an absence of explanation is fundamentally unsatisfactory, and James Hannam has cogently designated this work 'the least plausible creation account ever written'.[8] In coming sections, the theistic account of creation will be explored, beginning with the meaning of belief in God's transcendence, and in Chapter Five reasons will be given for adopting such an account.

Section 2: Transcendence

What is meant by the 'transcendence' of the God of theism should now be investigated. This transcendence consists in much more than a transcending of empirical verification (as in the issue of verification-transcendence discussed in Chapters One and Two). According to Owen, it consists, in part, in God's being substantially distinct from the world, a distinctness expressed by Karl Barth as God's 'qualitative difference' from humanity.[9] As H.D. Lewis puts this,

---

6 Owen, ibid., p. 75.

7 Peter Atkins, *The Creation*, p. 17.

8 James Hannam (2004), Review of Stephen Barr (2003), *Modern Science and Ancient Faith*, *Science and Religion Forum Reviews*, **43**, 16–18, p. 17. For a more detailed critique of Atkins's book, see Keith Ward, *God, Chance and Necessity*, Chapter 1.

9 Owen, op. cit., p. 35.

God's transcendence relates to God's 'beyondness' or 'otherness'.[10] The deity of pantheism would lack such otherness, as also would the finite deity of Platonism and of forms of panentheism for which God is dependent on the world. Only a God with the power, knowledge and wisdom of a creator is transcendent in this sense.

Pieter J. Huiser distinguishes three aspects of God's transcendence. The first is ontological; God is not identical with the universe, but completely independent of it. (This aspect overlaps with another aspect of Owen's account of transcendence, namely that God does not need the world.[11]) The second aspect is epistemological, an aspect expounded by Huiser as 'God does not coincide with the things we say about him';[12] given Huiser's commitment to religious realism, his point is perhaps that God's existence and nature are entirely independent of human conceptions of God, a theme implicit in the Old Testament passage of Isaiah, 'For my thoughts are not as your thoughts, neither are my ways your ways, says the Lord'.[13] Indeed the theistic conception is of a God whose existence and nature are independent of human conceptions (however this could itself be known). Huiser calls his third aspect 'the religious aspect' (better expressed, perhaps, as 'the axiological aspect', since the previous aspects are also relevant to religious attitudes), which he explains thus: 'God incorporates values, which transcend human ones'.[14] Without discussing what is here meant at length, we can take this to refer partly to God's perfection and partly to God as the creator of whatever has value (a theme returned to in Chapter Eight).

Huiser proceeds to discuss Phillips's account of divine transcendence. This he summarizes as having to do with values, like goodness and other-directedness, and with a religious anthropology, and not at all with epistemological or ontological issues. Transcendence of the world is construed as love which diverges from worldly self-centredness, and '"Other than the world" is taken to mean that the good is done out of love of the good only'.[15] But whatever we make of this radical, humanizing re-interpretation of talk of transcendence, Huiser's implicit criticism is that such an account altogether omits the ontological and epistemological aspects. Given that anti-realists are bound to reject the epistemological aspect as interpreted above, the central criticism concerns the anti-realist omission to include God's ontological independence within the meaning of 'transcendence'. Re-interpreted like this, transcendence becomes unrecognizable, for it is no longer God who is transcendent.

---

10 H.D. Lewis (1959), *Our Experience of God*, p. 66.

11 Owen, op. cit., p. 35.

12 Pieter J. Huiser (1997), *Models, Theories and Narratives*, p. 90.

13 Isaiah 55:8.

14 Huiser, op. cit., p. 90.

15 Huiser, ibid., p. 91.

Phillips has, however, more recently written of creation as renunciation, and this perspective on transcendence may seem more attractive.[16] His thinking here endorses that of Simone Weil, who contrasts creation with an act of imposing one's power. Creation, according to Weil,

> is an abdication. Through this act a kingdom was established other than the kingdom of God. The reality of this world is constituted by the mechanism of matter and the autonomy of rational creatures. It is a kingdom from which God has withdrawn. God, having renounced being its king, can enter it only as a beggar. As for the cause of this abdication, Plato expressed it thus: 'He was good'.[17]

These remarks suggest the possibility of reconciling belief in creation with belief both in evolution and in autonomous human development, and also convey that creation confers upon the world a kind of qualified independence, even though, according to theism, creatures remain dependent on the creator at all times. Weil's remarks also echo the ancient Platonic tradition (of the *Timaeus*) to the effect that if God had refrained from creating he or she would have shown grudgingness rather than goodness.[18] Weil's interpretation of creation, then, has value (although God cannot be held to withdraw from what did not exist but for God's bringing it into being). It should, however, be added that this interpretation of creation can only be adopted by those who recognize creation as the act of a creative agent, who exists independently of it. This suggests that those adopting this interpretation have to accept God's transcendence in the sense of his or her ontological independence.

The other aspect of God's transcendence mentioned by Owen is God's incomprehensibility. If this means that we have no grasp of how God creates, then it should be granted. However, if the suggestion is that we have no idea of God's nature, then it is a claim that undermines itself; for to know that God is incomprehensible, we would need to know a good deal about God. A parallel argument has been called 'the argument from imperfection'. Since human thinking is finite and imperfect (this argument runs), and God is infinite and perfect, all human conceptions of God must be inadequate and God must be incomprehensible to human beings.[19] But as Humphrey Palmer remarks in the same context, this argument has several flaws. Thus not everything finite need have inadequate ideas, even of what is infinite or perfect. More to the present point, the conclusion that God is incomprehensible to human beings conflicts with knowledge of the premise that God is infinite and perfect.[20] Further, if God's

---

16 D.Z. Phillips, *Religion and the Hermeneutics of Contemplation*, p. 100.

17 Simone Weil (1987), 'Are We Struggling for Justice?', trans. Marina Barabas, *Philosophical Investigations*, **10**.1, 1–10, p. 3.

18 Arthur O. Lovejoy (1936), *The Great Chain of Being*, pp. 47–9.

19 Humphrey Palmer (1973), *Analogy*, p. 26.

20 Palmer, ibid., pp. 26–8.

nature were incomprehensible, then the theory of analogy, which Owen endorses,[21] would not work, since we could not discover the level at which predicates can be ascribed to God.[22] As Keith Ward has written, 'It is ridiculous to say that God is wholly inconceivable, for that would leave the word 'God' without any meaning. There must be some things which are known about God, if we are to use the word intelligibly'.[23]

Owen, to his credit, rejects at least the extreme form of belief in God's incomprehensibility, adopted by Karl Barth. According to Barth, there is no similarity whatever between human creatures and the creator; nothing but being in receipt of God's grace makes it possible to speak meaningfully about God.[24] To this, Owen asks in reply how, if there is no analogy at all between God and his human creatures, God could reveal himself through a human being (as, of course, Barth held).[25] It could further be replied (although Barth would have rejected the reply) that there is nothing to prevent human beings asking intelligible cosmological questions and thus forming a conception of a deity with the power, knowledge and wisdom necessary to create the material world. It then becomes possible to talk of such a creator in the manner endorsed by the theory of analogy. But all this involves accepting that some attributes of humanity (power, knowledge and wisdom, for example) are applicable to God, and thus (to cite Owen's words) that 'there is a likeness, as well as an unlikeness, between God and his creatures'.

But this being so, God cannot be altogether incomprehensible. God's thoughts and ways may be vastly different from ours, and this difference may well be an aspect of the divine transcendence. But this qualitative unlikeness has to be qualified by the likeness presupposed if any of our talk of God is to avoid futility. To this, Owen adds further grounds; here are some of them. The creatures of a creator must bear some slight resemblance to their creator; for example, creatures as well as the creator have some kind of independent existence or autonomy. Further, from the perspective of theology, much the same claim is present in the Old Testament teaching (found in Genesis) that God created humanity 'in his own image'.[26] There is no need to adopt anthropocentric interpretations of this passage to find in it the belief that God, if God there is, and his or her human creatures have at least something in common; examples might include thought and speech, to say nothing of rationality. None of this detracts from God's transcendence, which fundamentally consists in God's independent existence and in the distinctive attributes that make creation possible.

---

21 Owen, p. 39. For the theory of analogy, see Chapter One, Section 6.

22 Robin Attfield (1971), 'The Individuality of God', *Sophia*, **10**.1, 20–27.

23 Keith Ward (1982), *Rational Theology and the Creativity of God*, p. 61. See also Alvin Plantinga (1980), *Does God Have a Nature?*

24 Owen, p. 39.

25 Owen, p. 39.

26 Owen, p. 40; Genesis 1:27.

Unlike anti-realists, there is nothing to prevent realists, however critical their realism, from accepting both these aspects of the meaning of divine transcendence.

Section 3: Creation, Necessity and Contingency

Could God have not existed? And is creation to be considered necessary or contingent? Most people accept that physical objects are each and all contingent, since each of them might have been otherwise and might not have existed at all, and also since it is a contingency that there are any such objects. Indeed the cosmological argument is often referred to as the argument from the contingency of the world. However, it is often claimed that God is in some sense necessary, whether because the same argument is thought to find an explanation of contingency in necessity, or because another argument, the ontological argument, maintains that the very concept of God is such that God's existence is necessary and God cannot not exist. Sometimes it has even been maintained that God's act of creation is also necessary, and even that the created world too could not have been otherwise, whether with regard to its contents, their histories, or the laws that govern them,[27] although this interpretation would of course preclude reasoning from its contingency. The relations of God and creation to necessity and contingency should accordingly be considered.

I shall suggest that it is best to avoid speaking of God as a necessary being altogether. The many writers who speak of God in this fashion usually have in mind not logically or conceptually necessary existence, but rather not being susceptible to creation, destruction or externally initiated change: or, in other words, existing independently, immutably and eternally. Patterson Brown, for example, has shown that the Third Way of Thomas Aquinas concerns not logically necessary existence, but (approximately) the kind of 'necessary existence' just mentioned.[28] For this, as has been argued above, is what believers in creation need to affirm about the creator; otherwise, cosmological questions raised about creatures could be raised about God as well. This kind of 'necessary existence' neither implies nor is implied by inability not to exist, as, for example, Owen accepts,[29] but since the two concepts are easily confused, it is best in discussions of creation to speak of God's independence and eternity instead. The existence of such a God is not a necessary truth, and belief in it is capable of being supported by grounds such as the cosmological argument purports to supply.

However, to go beyond this and claim that God's existence is a logically or conceptually necessary truth seems misguided. For one thing, this would imply

---

[27] Lovejoy, op. cit., depicts this historical view (which he rejects) at p. 328.

[28] Patterson Brown (1964), 'St. Thomas' Doctrine of Necessary Being', *Philosophical Review*, **73**, 76–90.

[29] Owen, op. cit., p. 15.

that God exists in every possible world, and therefore that it is inconceivable that there could have been nothing at all; yet we have no basis for making this assertion, and every reason to hold that a completely empty world is a conceivable possibility. There again, if God exists in every possible world, then his or her existence appears not to explain the world's contingency; for a necessary truth does nothing to explain anything that could have been otherwise.[30] Indeed, even if God's eternal and independent nature means that no explanatory questions about God make sense, nevertheless if we hold that God's existence is what explains the contingency of the material world, we also need to hold that it could have been otherwise (as many writers, including Terence Penelhum (a believer) and Robin Le Poidevin (an atheist), have recognized).[31] Hence discussions of creation need to avoid the suggestion that God's existence is logically or conceptually necessary. J.N. Findlay's ontological disproof of God's existence, which claims that nothing short of a logically necessary being is fit for worship, can thus be challenged,[32] since the universal creator is regarded by Muslims, Jews and Christians as fit for worship without usually being held to have logically necessary existence. Findlay's disproof thus collapses whether or not his other premise stands. His further premise is the questionable claim that there is nothing whose existence is logically necessary (not even realities, apparently); it is unnecessary, however, to question this here. Since a worshipful God need not have necessary existence, the claim that nothing has such existence would not in any case establish that there cannot be such a God.

The traditional ontological argument, by contrast, purports to show that God cannot not exist, and claims to do this simply on the basis of one or another definition of God. Many attempts have been made to show that no such argument can succeed, but these attempts have often involved recourse to contestable claims such as that 'existence' is not a predicate, and therefore cannot be included in definitions and/or descriptions. I find such claims unpersuasive, since 'existing' (and cognate terms such as 'historical') are often used, in expressions such as 'the existing authorities' or 'historical event' to convey current or past actuality. But Jerome Shaffer seems correct in holding that, besides this 'intensional' sense of 'exist(s)', there is an 'extensional' sense conveying that one or another concept has application, or is instantiated, and that 'exist(s)' in this sense cannot be included in any definition, since otherwise we can define things into existence. But this extensional sense is the sense that any successful ontological argument

---

30 William Rowe, 'The Cosmological Argument', pp. 58–9; Ward, op. cit., p. 8; Robin Attfield (1983), 'Necessity and Contingency in God', *New Blackfriars*, **65**, 35–41.

31 Terence Penelhum (1960), 'Divine Necessity', *Mind*, **69**, 175–86, p. 185; Robin Le Poidevin (1996), *Arguing for Atheism*, pp. 40–41.

32 J.N. Findlay (1955), 'Can God's Existence be Disproved?', in Flew and MacIntyre, *New Essays*, 47–56.

would need to employ to show that God exists.[33] Accordingly all forms of the ontological argument miscarry.

Hence the view of Immanuel Kant that the Cosmological Argument turns on the Ontological Argument[34] turns out to be an overgeneralization; even if this was true of some eighteenth-century versions, it cannot be true of all versions, since the creator of the Cosmological Argument need not and should not be conceived as necessarily existent. Equally important, the central objections of David Hume to the Cosmological Argument, concerning as they do belief in a deity who cannot not exist, turn out to be beside the point.[35] Hume had further objections, to be considered in the next chapter. But the apparent discomfiture of his character Demea, the defender of this Argument, misleadingly suggests that it falls simply because of objections such as that on which Hume's Cleanthes claims to 'rest the whole controversy', namely that 'There is no being, therefore, whose non-existence implies a contradiction'.[36] But if the whole controversy were to rest on this objection, the argument would remain unscathed.

If we discard the concept that God cannot not exist, and with it the view that everything true of God is logically necessary, an important motivation disappears for holding that the act of creation is itself a necessary truth. Some pantheistic systems, such as that of Spinoza, have combined belief in God's necessary existence with the supposed necessity of the material world being as it is, and of the relation of God and the world being as it is conceived to be in such a system; but systems of this kind turn out to be untenable, since they leave no room for contingency at all. Certainly, if we argue from God's nature to the existence of the world as some kind of necessary and necessitated implication, as neo-Platonists have usually done, then belief in the necessity of the world and of its relation to God becomes understandable.[37] Yet nothing in the concept of creation, nor in arguments from the world to God such as the Cosmological Argument, supports such inferences, which cannot but be regarded as groundless. Besides, it has already been remarked that necessary truths cannot explain contingencies; thus if we treat the contingency of the material world as given, we have an extra reason to resist the view that God's act of creation would be necessary. As Ward (who argues along similar lines) remarks, 'a necessary being cannot give rise to a world of contingent, free creatures'.[38] Further, if we were to discard the view that physical

---

33 Jerome Shaffer (1962), 'Existence, Predication and the Ontological Argument', *Mind*, **71**, 307–25, pp. 322–5.

34 Immanuel Kant [1781] (1968), *Critique of Pure Reason*, Transcendental Dialectic, Book II, Chapter iii, N. Kemp Smith (trans.), *Immanuel Kant's Critique of Pure Reason*, pp. 485–531.

35 David Hume [1779] (1963), 'Dialogues concerning Natural Religion', Section IX, in Richard Wollheim (ed.), *Hume on Religion*, 99–204, pp. 162–5.

36 Hume, ibid., pp. 162–3.

37 Owen, op. cit., pp. 59–65.

38 Ward, op. cit., p. 215.

objects (whether severally or jointly) could have been otherwise, then we seem to have discarded the very concept of contingency which allows issues of contingency and necessity to arise and be discussed. Thus neither God's act of creation nor God's existence are to be regarded as logically or conceptually necessary.

If the material world is recognized to be contingent, this also applies to the laws of nature, which could have been otherwise. Certainly they can also be regarded as specifying what is physically possible or necessary; but these physical possibilities and necessities are conditional on the laws being as they are. Yet we can conceive of worlds in which, instead of inverse square laws of gravity, light diffusion and the like, there could have been inverse cube laws. Everything would have been almost unimaginably different, but that does not make such worlds logically impossible. And if there is a creator, then the laws that actually prevail are best understood as expressing his or her will. There would be no need to postulate a subordinate deity (such as Plato's demiurge) to explain the kind of order that science studies, but there would be every reason to ascribe the regularities of nature, as well as the existence of the material world, to the same creator.

Section 4: Creation and Time

The relation of God and of creation to time must now be considered. Belief in creation involves belief in the dependence of physical objects on God. But physical objects are spread out across time. Hence belief in creation concerns the dependence of physical objects at all times. It does not only concern the origin of such objects (and thus, possibly, the moment of the Big Bang). Hence, as was argued in Chapter One, the belief that the world's past is finite is not essential to belief in God as creator, although it is compatible with that belief. While some physicists say that the Big Bang was the earliest event, others deny this, and postulate a prior contraction, preceded possibly by a yet earlier expansion. Belief in creation is not committed to either of these views, but is consistent with each of them. Thus even if there was no earliest event, and the past was not finite, belief in creation remains unaffected.

In any case, God's bringing about of physical objects could not be an event prior either to the Big Bang (if that was the earliest event) or to the sequence of physical events that preceded it from earliest times (if there was such a sequence). For there cannot have been an event prior to the earliest event, and if the sequence of physical events that may have preceded the Big Bang comprised the earliest events, there cannot have been an event prior to these earliest events. Possibly God's creative action should be seen as inaugurating time, but if so this action would not itself be an event at all, but (in the words of Michael Durrant) would set 'a limit' to time.[39] Yet even if there has been a sequence of physical events

---

[39] Michael Durrant (1973), *The Logical Status of 'God'*, pp. 23–4.

stretching infinitely into the past, creation turns not on the extent of the past but on whether or not the whole sequence is and has been dependent on God.

This becomes clearer if a distinction of Peter Geach is remarked. While creation can be regarded as a kind of causing, it should not be understood on the model of one event causing another event. In some respects, talk of divine creation is much more similar to 'Phidias caused the block of marble to be of human form', except that Phidias was effecting a change in something already existent (the block of marble), whereas God is to be understood not as modifying anything existent but as bringing creatures into existence which would not have existed otherwise.[40] Yet the similarities are instructive. For the statue went on having a human form for many years, and thus at times quite different from that of Phidias' act. Bringing about, then, need not be regarded as simultaneous with states or events brought about. Language about bringing about is liable to involve two time-references, that of the action and that (or those) of the ensuing states and events. God's act of creation, then, need not be regarded as simultaneous with the objects or events or states brought into being, even if it is or was situated in time at all.

Indeed, as has been argued already, God, as creator, would have to be immune from the kind of change that his or her own creative activity would generate and explain, and, for parallel reasons, unable to be brought into being or destroyed. For otherwise God would be too similar to creatures, and in relevant ways in need of explanation just as they are. Hence God has to be regarded as immutable, uncreatable and indestructible. (Process theologians, in suggesting otherwise, adopt a stance incompatible with belief in creation, and also seem to block the possibility of cosmological reasoning from creatures to their creator. For if God were dependent on the world or subject to change, we would need an explanation of divine dependence as well as of creaturely dependence, and could not explain creaturely dependence by reference to an independent creator.) Such a God can still be regarded as sustaining the existence of the material world; indeed if its dependence on God at all times is accepted, then divine preservation becomes continuous with divine creation,[41] the difference being that creation involves bringing creatures into being, while preservation concerns keeping them in being. But should God's creative activity therefore be understood to be performed at all times? This does not follow from the world's dependence on God at all times, since, as was argued above, the time-reference of creation need not be the same as that of the states of creatures generated, or the events affecting them.

Some believers in creation nevertheless regard God as acting within (indeed throughout) time. If so, God, like his or her creatures, has duration, but in the case of God that duration is everlasting or 'sempiternal'. But this interpretation makes God subject to temporal limitations, and also implies that it makes sense to ask about God how long he or she has existed; and yet, as Durrant has remarked,

---

[40] Peter Geach, *God and the Soul*, pp. 82–3.

[41] Owen, op. cit., p. 10.

this is surely a nonsensical question.[42] Another possible implication is that the decision to make the laws of nature applicable across the future has not yet been taken, as this form of the dependence of creatures would probably have to be regarded as decided for each future time at that time. Yet believers in God's constancy hold that this belief implies that the future will embody regularities corresponding to those of the past and of the present; and they do not believe that God could change his or her mind about this.

It is therefore preferable, I suggest, to regard God's existence and activity as not situated in time, but as timeless, as was held by Augustine, Boethius, Anselm and Aquinas.[43] If so, God timelessly brings it about that physical objects exist at particular times, and comply across time with diachronic regularities or laws of nature. God's incorporeality is not usually considered a problem for belief in creation, given that a corporeal (and thus spatial) being could not create physical objects, being one of them. Similarly, it may be held, God's atemporality presents no greater problem, when the creator of temporal objects is in question. Indeed if God is responsible for whatever exists in space and time, then his or her existence and activity must apparently be independent of time as well as of space; for a temporal God would be too similar to the other inhabitants of time. God's atemporality is compatible with the block-universe theory of time, for which the past, the present and the future comprise one existent manifold or continuum, but equally with the process or tensed view of time, for which only the present exists and the future is as yet non-existent.[44] God's omniscience can also be understood as timeless, and as cohering with either of these theories, despite objections shortly to be mentioned.

Objections to belief in a timeless creator have been presented by J.R. Lucas, Nelson Pike and Anthony Kenny. Lucas objects that on this view God is insufficiently personal, since to be a person one must be capable of consciousness, and all consciousness is temporal.[45] However, if there is a conscious creator of temporal beings who is non-temporal, then not all consciousness can be temporal. Pike follows Friedrich Schleiermacher in doubting whether a timeless God could create, since creation supposedly involves a temporal relation between the creator and the created;[46] but this supposition ignores the double time-reference of 'bring about' and of 'create' remarked by Geach. For his part Kenny takes up the ancient suggestion that timeless knowledge would make God aware of all temporal events

[42] Durrant, op. cit., p. 24.

[43] Brian Davies (1983), 'A Timeless God', *New Blackfriars*, **64**, 215–24, p. 215.

[44] Paul Copan and William Lane Craig are among those accepting both God's timelessness and the process or tensed view of time; see Copan and Craig, *Creation out of Nothing*, p. 253.

[45] J.R. Lucas (1970), *The Freedom of the Will*, p. 75. William Lane Craig well replies to criticisms of divine timelessness from Lucas and others in 'Divine Timelessness and Personhood' (1998), *International Journal for the Philosophy of Religion*, **43**, 109–24.

[46] Nelson Pike (1970), *God and Timelessness*, pp. 109–10, 189–90.

together (*totum simul*), in the manner of a contemporary observer, as implying that each event (past, present or future) becomes simultaneous with God's knowledge, and thus, by transitivity, with every other event.[47] But as Brian Davies has replied, this objection ignores the fact that such knowledge is not held to be contemporary or simultaneous with its objects at all; rather God is held to see all truths (past, present and future) together,[48] somewhat like a composer who contemplates the whole sequence of a musical composition without the contemplation having to be situated within the duration of the musical sequence.

There are thus, as we have seen, reasons to prefer the timelessness view, which is not defeated by at least its most prominent objectors. This view has been held to cohere well with the contemporary cosmological insights of physical science, and thus to capture God's transcendence over space-time,[49] and also brings the asset that the apparently insoluble problem of reconciling God's knowledge with free human actions of the future becomes capable of solution, as suggested long ago by Boethius.[50] If God has necessary (and comprehensive) foreknowledge, then the humans whose actions are foreknown apparently cannot be free. But if God's knowledge is timeless, then these actions are not foreknown at all, and, as far as God's knowledge is concerned, may be just as free as free human actions that are known through the observation of contemporary observers. Given the timelessness view, then God's knowledge will extend to whatever truths can be known in an atemporal or timeless manner, including whatever truths about the future are knowable; and knowledge of this kind will also be the context of God's powers.

The sense in which God is said to be omnipotent can now be introduced. Here I shall use the definition of Bruce Reichenbach:

> A being *x* is omnipotent if and only if it is capable of bringing about any contingent state of affairs (a) whose description does not contain or entail a contradiction and (b) whose description does not exclude or entail the exclusion of *x* or any omnipotent agent from among those which may have brought about that state of affairs.[51]

This definition avoids the pitfalls of endowing omnipotence with the ability to do what is contradictory, and the so-called 'paradoxes of omnipotence' as well. It also coheres with God's timelessness, and simultaneously captures the kind of power that must be ascribed to a creator.

Thus the relation of creation to time can be understood on the basis of timeless creative activity bringing it about that physical objects, their states and the events

---

47 Anthony Kenny (1979), *The God of the Philosophers*, 38–9.

48 Brian Davies (1982), 'Kenny on God', *Philosophy*, **57**,105–17, pp. 105–6; see also Davies, 'A Timeless God'.

49 Willem B. Drees (1996), *Religion, Science and Naturalism*, p. 266.

50 Boethius (1897), *De Consolatione Philosophiae*, trans. George Colville, ed. Ernest Belfort Bax, V, vi, ll. 63ff.; quoted by Lucas, op. cit., p. 73, n.3.

51 Bruce Reichenbach (1982), *Evil and a Good God*, p. 44.

that befall them exist or happen at temporal moments or for temporal durations, and thus making the experience of creatures possible. This activity also brings it about that physical objects comply with lawlike regularities that have application at all times and places.

Section 5: Creation without Creationism?

In a work concerning connections between creation and Darwinism, consideration is in place about the relations between belief in creation, as discussed in this and previous chapters, and creationism, usually regarded as an anti-Darwinian and generally anti-evolution movement or position. Need believers in creation be creationists? Tackling this question involves eliciting just what creationism involves.

There turn out to be not one but two positions known by this name, both of which turn on belief in the special creation by God of all existing natural species. The first of these positions is that of nineteenth-century theistic opponents either of evolution in general or of Darwin's theory of evolution by natural selection. This widely held position often comprised an approach to science for which science and religion were inseparable and the methodology of science was to embody this inseparability.[52] Accordingly, Neal C. Gillespie has labelled creationism an 'episteme', to be contrasted with the rival episteme of positivism.[53] He borrows the term 'episteme' from Michel Foucault, according to whom an episteme is 'the historical *a priori*' that

> in a given period, delimits in the totality of experience a field of knowledge, defines the mode of being of the objects that appear in that field, provides man's everyday perception with theoretical powers, and defines the conditions in which we can sustain a discourse about things that is recognized to be true.[54]

Gillespie's own account of these two epistemes runs as follows:

> The positivist limited scientific knowledge, which he saw as the only valid form of knowledge, to the laws of nature and to processes involving 'secondary', or natural causes exclusively. The creationist, on the other hand, saw the world and everything in it as being the result of direct or indirect divine activity. His science was inseparable from his theology. ... To comprehend more fully, for a scientist of this persuasion, was to understand the workings of the mind of the Creator.[55]

---

52 Neal C. Gillespie (1979), *Charles Darwin and the Problem of Creation*, pp. 13–16.

53 Gillespie, ibid., pp. 1–2.

54 Michel Foucault (1972), *The Archaeology of Knowledge*, p. xxii; Gillespie, ibid., p. 2.

55 Gillespie, ibid., p. 3.

Gillespie sees Darwin as increasingly embracing the positivist episteme and introducing it into biology, without ever completely shaking off creationism.

But these distinctions between discourses and fields of knowledge seem at once too demanding and somewhat arbitrary. Thus at least the first sentence of Gillespie's characterization of creationism fits Christian followers of Darwin such as Asa Gray, and thus seems to define creationism too broadly; for those evolutionary theists who did not regard the species as special creations still regarded them as created indirectly. If, however, the last sentence is pressed, and nothing would be counted by a creationist as a gain to understanding unless it referred to the divine mind, then few of the naturalists among the Victorian clergy will count as creationists, in view of their widespread fascination with humdrum gains to knowledge in the fields of habitats, colonization and classification. Correspondingly the above account of positivism excludes everyone who recognized non-scientific fields of knowledge such as mathematics and history, and is thus similarly over-demanding. To represent these two stances as paradigms between which (if we follow through the implications of Foucault's definition) there could be little or no rational debate involves a misleading idealization of the debates that actually took place.[56] Gillespie certainly admits that his characterizations of these epistemes embody broad-brush approximations,[57] and as such they have value; but his characterization of creationism in particular cannot be reliably employed to chart the relations of anti-Darwinist stances to belief in creation. Rather, nineteenth-century creationism should be understood as involving belief in the special creation of natural species and as tending to involve belief in the inseparability of science and religion, as expressed by creationists such as Adam Sedgwick[58] and the Duke of Argyll.[59] So in one form the question with which this section opened becomes that of whether believers in creation need accept all (or any) of this.

The second position known as 'creationism' is a phenomenon of the twentieth and twenty-first centuries, and has been consciously devised as a rival to Darwinism. Allied to creationism has been a would-be scientific position concerning the origins of species, creation science, which seeks to explain these origins along the lines of the text of Genesis. However, in a court case in Arkansas in 1981, the judge rejected the claims of creation science to be science, and to be taught as such in state classrooms, accepting the testimony of the philosopher of science and Darwinian Michael Ruse, himself sympathetic to Christianity, that creation science, being unfalsifiable, is not science, but a variety of religion.[60] Meanwhile, theistic opponents of Darwinism (some of them Islamic, but most

---

56 For the notion of incommensurable paradigms, see Thomas S. Kuhn (1962), *The Structure of Scientific Revolutions*.

57 Gillespie, ibid., pp. 1–3.

58 Gillespie, ibid., pp. 14–15.

59 Gillespie, ibid., pp. 14, 93–104.

60 Michael Ruse (2001) *Can a Darwinian Be a Christian?*, pp. 4, 6–7.

of them Christian) have contested the ability of Darwinism to explain, for example, the origin of life.[61] These opponents have included the distinguished philosopher and epistemologist Alvin Plantinga.[62] Another prominent Christian opponent of Darwinian explanations is William A. Dembski, who argues that Darwinism fails to explain the development of life on earth, and that the only alternative is what Dembski calls 'Intelligent Design'.[63] Yet another is Michael Behe, whose book *Darwin's Black Box* presents an empirical case for conscious design in biology; the world revealed by biochemistry is supposedly too irreducibly complex to be explained by gradualist Darwinian explanations.[64] The work of this group of writers (Islamic as well as Christian) is sometimes known as 'the New Creationism' on the part of opponents such as Ruse. They would probably not be regarded (even by Gillespie) as sharing a paradigm or discourse or episteme. However, twentieth and twenty-first century anti-Darwinian creationists share with their nineneenth-century pre-Darwinian and late Victorian predecessors belief in the special creation of (at least some) natural species and a tendency to regard science and religion as needing to be studied in tandem. Thus our opening question becomes that of whether believers in creation need share this belief and this tendency.

With all this emphasis from advocates of creationism on belief in creation, there is a tendency for intelligent members of the public (philosophers included) to conflate or confuse these beliefs. Indeed, there is nothing to prevent creationists being committed to the dependence of physical objects at all times on God as creator. Most, but not quite all creationists are so committed; the exceptions are people impressed with arguments for design who postulate a cosmic designer not regarded as the world's creator,[65] rather along the lines of Plato's demiurge, or of the limited and dependent deity of process theology. Some creationists, it should be added, envisage God as acting at one moment or a series of moments in time to create a succession of species; however, creationists are also free to adopt the timelessness view of creation, and claim that God's timeless creation brings species into existence at this succession of times. Either way, they are committed to supernatural interventions of a miraculous nature, yet not necessarily to representing God as creating through a series of afterthoughts.

---

[61] Ruse, ibid., pp. 60–66.

[62] Alvin Plantinga (1991), 'When Faith and Reason Clash: Evolution and the Bible', *Christian Scholar's Review*, **21**.1, 8–32; reprinted in David L. Hull and Michael Ruse (eds) (1998), *The Philosophy of Biology*, 674–97.

[63] William A. Dembski (1998), *The Design Inference*; Dembski (2003), *The Design Revolution*.

[64] Michael Behe (1996), *Darwin's Black Box*; Behe (2004), 'Irreducible Complexity: Obstacle to Darwinian Evolution', in William A. Dembski and Michael Ruse (eds), *Debating Design*, 352–70.

[65] Behe is an example. See Ruse, op. cit., pp. 119, 120.

But believers in creation (as depicted in earlier sections of this chapter) are not in general committed to creationism, whether because they endorse Darwinism in general or in one or another of its contemporary varieties, or because they have no views on the explanation of the existence of species, or because they regard God's purposes as unknowable. Nor need they harbour any such commitment. There is, after all, nothing to prevent believers in creation from adhering to a Baconian belief in the methodological separation of science and religion (a position dubbed 'methodological naturalism' in Chapter Six), and/or rejecting the search for irreducible 'final causes' or purposes in biology. A person can believe in the dependence of the material world on God, without holding that God intervenes in the course of time to bring into being species that would not have existed otherwise. And this answers our opening question: creationism is not implicit in belief in creation.

Michael Ruse lists a number of prominent Darwinians who have also been theistic believers, including in the twentieth century Ronald Fisher (an Anglican) and Theodosius Dobzhansky (of the Russian Orthodox Church), both of them prominent in advocating the synthesis of Darwinian natural selection and Mendelian genetics that entrenched modern Darwinism in the thirties and forties.[66] Nor are these leading scientists isolated exceptions; for the main Protestant churches accepted Darwinism during the later decades of the nineteenth century, followed some decades later (and more reluctantly) by the Roman Catholic Church.[67] Creationism is thus a minority movement among Christians, despite its widespread (and possibly growing) following in the United States. Indeed many Christians, Jews and Muslims believe that God does not interrupt the regular sequences of nature, but governs entirely through laws of nature that apply across space and time. (This belief is sometimes belaboured with charges of 'deism'. By this phrase, its critics tend to mean, albeit unhistorically, the theological view that God started off the cosmic process and then ceased to play any further part therein. But that particular view of divine action and of the relation of God to time is by no means implicit in the belief that God chooses to govern by laws rather than by supernatural interventions. Indeed holders of this belief may be more consistent believers in God's wisdom and providence than those who think that without miracles God's purposes would be frustrated; for the laws of nature could be held to embody God's original intentions without any need for corrective activity.) As will be seen in the next section, this same belief was (for at least two decades) the stance of Charles Darwin himself.

---

[66] Ruse, op. cit., pp. 8–9, 26.

[67] For Protestant denominations, see Owen Chadwick (1966), 'Evolution and the Churches', in Owen Chadwick, *The Victorian Church*, Part II, 23–35; reprinted in C.A. Russell (ed.) (1973), *Science and Religious Belief*, 282–93. For papal pronouncements, see Stephen Jay Gould (2002), *Rocks of Ages*, pp. 75–82.

Section 6: Darwin on Creation

Charles Darwin (1809–1882), the originator of the theory of evolution by natural selection,[68] was for most of his life a believer in God as creator, but in his later years became an agnostic. However, for more than half of his life, the part beginning in 1838 (when he read Malthus and lighted upon natural selection as the main mechanism of evolution), he was critical of the theory of the origin of species that invoked special creation.[69] All this indicates that for a long period during the middle of his life (for well over two decades from 1838) Darwin combined rejection of special creation with acceptance of creation on the part of the God of Christian theism. How Darwin reconciled these beliefs but later came to discard belief in creation is the theme of this section. Darwin's intellectual development serves as a highly apposite historical case-study, not least through illustrating that this combination of beliefs is possible, but also through foregrounding many of the issues (adaptations, design arguments, parasitism and providence included) discussed more systematically in the four chapters that follow.

The nature and sequence of Darwin's religious beliefs has been ably charted by Maurice Mandelbaum,[70] whose basic account is followed here; some further insights have been added by John Hedley Brooke, particularly on Darwin's later loss of religious belief, and by Robert J. Richards.[71] We should begin with a work of Darwin, probably of late 1838, with the pertinent title 'Essay on Theology and Natural Selection',[72] although it is hardly an essay, but a set of critical notes on a creationist work of the day.

Here Darwin rejects the suggestion that the transportation of seeds through the air and of coconuts by water is directly designed, but does 'not want to deny laws', recognizes that 'the whole universe is full of adaptations', but holds that these are 'direct consequences of still higher laws'. Thus 'the laws of transportation were created with reference to successive development' (of species); but Darwin adds, characteristically, that this admission is probably based on his own 'ignorance'.[73] Darwin here suggests, then, that the laws of nature could have been so devised

---

[68] Having discovered natural selection at least as early as 1838, Darwin must be regarded as anticipating the independent discovery of the same theory by his younger contemporary, Alfred Russel Wallace (1823–1913).

[69] Howard E. Gruber (1974), *Darwin on Man*; see Chronology, xxiii–xxv.

[70] Maurice Mandelbaum (1958), 'Darwin's Religious Views', *Journal of the History of Ideas*, **19**.3, 363–78.

[71] John Hedley Brooke (1985), 'The Relations Between Darwin's Science and His Religion', in John Durant (ed.), *Darwinism and Divinity*, 40–75; and Brooke (2002), 'Revisiting Darwin on Order and Design', in Niels Henrik Gregerson and Ulf Görman (eds.), *Design and Disorder*, 31–52; Robert J. Richards (2002), *The Romantic Conception of Life*.

[72] Gruber, op. cit., 414–22; cited by Brooke, 'Darwin's Science and His Religion', p. 47.

[73] Gruber, ibid., p. 416; Brooke, ibid., p. 47.

that the various species would develop from one another over the course of time. This was not his eventual view, but it remains a possible view of the relation of creation to the origination of species. As he wrote in the first transmutation notebook (begun 1837), 'the creator creates ... by laws'.[74] (Darwin used the term 'transmutation' of the evolution of new species from earlier species.) But as he comments in the 'Essay on Theology' (about the plants of deltas), 'the plants were no more created to arrest the earth' [sc. from being washed into the sea] 'than the earth revolves to form rain to wash down earth from the mountains upheaved by volcanic forces, for these marsh plants. All flow from some grand and simple laws.'[75] And when the creationist author he is criticizing (Macculloch) claims that the existence of useless bones is due to the creator's persistent pursuit of a design until it is exhausted and 'must be abandoned for another', Darwin ironically remarks: 'the designs of an omnipotent creator, exhausted and abandoned', adding: 'Such is man's philosophy, when he argues about his Creator!'[76] This is a rejection of contemporary natural theology, made for the sake of a more fitting grasp of creation.

During the same period (1836–1839) Darwin, who had at one time held fully orthodox Christian beliefs, abandoned belief in miracles and revelation.[77] Mandelbaum suggests that what undermined his orthodoxy was 'his acceptance of Lyell's views in geology, which ran counter to the orthodox interpretation of the creation'. Charles Lyell (1797–1875) adhered to the methodological stance of uniformitarianism, according to which nothing but the laws currently in force are to be adduced in scientific explanations.[78] Such an approach could well undermine both a belief in miracles and also an interventionist account of the origin of the species. It implied instead gradual change, happening over extremely long periods of geological time, implications now endorsed by Darwin. However, according to Mandelbaum, it was only after 1838, when Darwin had hit upon the principle of natural selection, that the argument of design put forward by William Paley 'lost all its force for him', since the biological adaptations of creatures to their environments and manner of life, to which Paley had appealed, could now be explained without resort to purposive supernatural agency of the kind implicit in creationist thinking.[79] (The general bearing of this on arguments from design will be considered in Chapters Five and Six.)

---

[74] Gavin de Beer (1960), 'Darwin's Notebooks on the Transmutation of Species', *Bulletin of the British Museum (Natural History)*, Historical Series, **2**, parts 2–5, p. 53; Brooke, ibid., p. 46. Robert J. Richards shows that at least in the 1840s Darwin associated natural selection with divine forethought: see *The Romantic Conception of Life*, p. 536.

[75] Gruber, op. cit., p. 417.

[76] Gruber, ibid., p. 417.

[77] Mandelbaum, op. cit., p. 364.

[78] See Charles Lyell [1830] (1875), *Principles of Geology* (12th edn).; also Charles Coulston Gillispie (1959), *Genesis and Geology*, 121–48.

[79] Mandelbaum, op. cit., p. 36.

Nevertheless, when Darwin published *The Origin of Species* in 1859, he continued to believe (as in earlier essays of 1842 and 1844) in the consistency of evolution by natural selection with theistic belief in creation. This, as he recognized, was because his work was dealing with the realm of 'secondary causes', and in no way involved metaphysical theories nor clashed with belief in God as first cause.[80] More positively, he welcomed the comment on his theory of the author, botanist and clergyman Charles Kingsley sufficiently to include it in the second edition of 1860 and in subsequent editions:

> I have gradually learned to see that it is just as noble a conception of Deity, to believe that he created primal forms capable of self-development into all forms needful *pro tempore* [for their time] and *pro loco* [for their place], as to believe that he required a fresh act of intervention to supply the lacunas which he himself had made. I question whether the former be not the loftier thought.[81]

Consistently, many of Darwin's criticisms of belief in special creation have no bearing on theistic belief in general, but stress, for example, the implicit appeal of the creationists to whole strings of miraculous interventions.

However, as Mandelbaum remarks, some of Darwin's criticisms could possibly be seen as carrying adverse implications for theistic belief. Thus he could not believe that natural variations, such as the tail of the woodpecker, were providentially arranged, and this view could be seen as undercutting one kind of argument from design. However, advocates of design arguments need not appeal to specific adaptations, as opposed to laws of nature (see Chapters Five and Six below). Again, Darwin's view of adaptations was that they were beneficial to the bearer's species only, and this was capable of generating theological problems. However, if the existence and flourishing of the bearer species is valuable, then this kind of distinctive beneficiality is not of itself a problem, as opposed to premature deaths and sometimes extinction among other species (topics to which we return in Chapters Six and Seven). Thirdly, Darwin found it difficult to believe in the creation of parasites, whether by special creation (his immediate target), or implicitly in any other manner consistent with divine purposiveness (another topic to be returned to).[82] Nevertheless at this stage Darwin continued to adhere to the view that the wonders of nature and the capacities of humanity could not have arisen through 'blind change or necessity';[83] his science and his religious beliefs remained cotenable.

Subsequently, however, Darwin was to express dissatisfaction with the view that design lay in general laws rather than in their detailed application; such an approach seemed to sidestep the problem of natural evil (evils not resulting

---

80 Mandelbaum, ibid., pp. 366–7.
81 Mandelbaum, ibid., p. 368.
82 Mandelbaum, ibid., pp. 369–70.
83 Mandelbaum, ibid., p. 371.

from human choices, that is: we return to this topic in Chapter Six).[84] He also increasingly took the view that the human mind is inadequate to appraise issues such as that of cosmic design.[85] Thus in his later years, concluding that there were no positive grounds for religious belief, Darwin became a fairly unqualified agnostic, and seems to have held back from atheism mainly from a distaste for dogmatism.[86] Yet his agnosticism was far from resolute: as he wrote when preparing his autobiography in 1879, 'My judgement often fluctuates … In my most extreme fluctuations I have never been an Atheist in the sense of denying God. I think that generally (and more and more as I grow older), but not always, that an Agnostic would be the more correct description of my state of mind.'[87]

Brooke argues convincingly that Darwin's change of beliefs is not to be attributed to commitment to natural selection.[88] Thus Darwin had held that commitment since 1838, whereas this change took place during the 1860s. Nor is his awareness of the scepticism of Hume and of the positivism of Comte likely to have caused this change; once again, he had been aware of them since at least 1838.[89] The explanation is rather to be found in several mutually reinforcing factors, including personal tragedies, an intense sensitivity to pain and suffering, a growing conviction that the difference between the mind of a dog and the human mind (and thus the human capacity to cope with cosmic issues) is one of degree, and Darwin's consciousness of his inability to explain particular disasters in providentialist terms. To these factors it can be added that the concern of his wife that he should share her religious convictions, together with the advocacy by friends of a stronger purposive content to biological laws, made him face up to doubts that might otherwise have remained dormant.[90] While his science drew his attention to some of these factors, it was not strictly this that required his change of mind, either logically or psychologically.[91]

Besides presenting an example of the possibility of adherence to theism and creation alongside rejection of creationism and special creation (temporary as such an adherence proved in Darwin's case), the life and intellectual development of Darwin draw attention to the importance of arguments for theism and of problems faced by these arguments, and also of possible tensions between theism and Darwinism. Key theistic arguments are considered in the coming chapter,

---

84 Mandelbaum, ibid., p. 372.

85 Mandelbaum, ibid., pp. 374–5.

86 Mandelbaum, ibid., pp. 376–8.

87 Alister McGrath (2005), *Dawkins' God*, p. 76. Stephen Jay Gould takes insufficient account of this passage of Darwin at *Rocks of Ages*, pp. 33–4.

88 Brooke, 'Darwin's Science and His Religion', pp. 59–61; 'Revisiting Darwin', pp. 38–41.

89 Brooke, 'Darwin's Science and His Religion', p. 61.

90 Brooke, ibid., pp. 62–9.

91 Brooke, ibid., pp. 58–61.

and key objections to them, as well as the bearing of Darwinism on theism, in that chapter and the next.

Chapter Five

Arguments from World to God

Section 1: From World to God

As we have seen in Chapters One and Two, questions about what explains the existence of the material world are not obviously incoherent. They need not involve attempts to explain everything that exists by reference to something beyond everything. Rather, they seek an explanation of the set of material or physical objects by reference to something with the power to bring them into being. As has also been seen in Chapter Four, the concept of creation concerns just such a relation between creatures and creator.

This chapter concerns grounds for belief in the God of theism. Such a God has to be understood as an individual whose agency 'makes a world of difference' (Chapter Three, Section 3). Anti-realists reject this understanding of God; hence it is unsurprising that Phillips regards arguments from world to God as incoherent, since the existence of God is out of the question in the first place.[1] But if the concept of God is of an individual capable of bringing the material world into being, it will be important to investigate whether there are grounds for belief in such a God, grounds based either on there being a material world or on some of its characteristics. Indeed, as Hepburn suggests, reasoning such as the Cosmological Argument from the world to God is plausibly indispensable to any defence of belief in God,[2] whether Christian (as in Hepburn's case), or of any other theistic kind. For arguments from particular characteristics need get no further than an agency responsible for those characteristics, but arguments from there being a world rather than nothing point, if valid, to an all-powerful creator.

The pivotal role of the Cosmological Argument seems to have been entirely missed in John Dupré's recent book *Darwin's Legacy: What Evolution Means Today*. Dupré argues that Darwin's theory of natural selection removes the need for a creator to explain the characteristic of apparent design, and regards this as the only rational basis for theism. Hence a large part of Darwin's legacy supposedly consists in removing the last support for belief in a God.[3] As we have seen in Chapter Four, Section 6, and as Dupré recognizes, this was hardly Darwin's own view; later I shall be arguing that acceptance of natural selection does not in any case undermine all versions of the design argument. But the present

---

1 D.Z. Phillips, *Faith and Philosophical Enquiry*, pp. 85, 130–31.

2 R.W. Hepburn, 'From World to God', p. 43.

3 John Dupré (2003), *Darwin's Legacy*.

point is that Dupré writes as if the Cosmological Argument had no bearing at all on grounds for belief in God, let alone comprising their core. That is why I am discussing and defending it first in this chapter (in the first two sections), and versions of the design argument afterwards (in the rest of the chapter). While both kinds of argument can be held to support the same overall theistic conclusion, the Cosmological Argument alone is directly relevant to God's omnipotence, independence and uncreatability, and to the dependence on God of the material world.

Some theists consider the Ontological Argument as a basis for theistic belief. Indeed Phillips finds value in this argument, partly because God is not portrayed as an individual. However, as was argued in Chapter Four, Section 3, that argument itself fails, and the related concept of God as necessarily existent is misguided. Indeed theists believe, as Terence Penelhum remarks, that if there were no God there would be nothing,[4] and thus do not believe that God exists in every possible scenario. It is rather the Cosmological Argument that reasons to God as creator and, as such, as an individual, if it can avoid the pitfalls with which it has been charged.

For Phillips and others regard such arguments as incoherent not only because of their objections to talk of God's individuality and existence but also because in their view attempts to treat the world as one big thing in need of explanation fail.[5] Replies to Phillips's objections have already been presented in Chapter Two, Section 6, but some related pitfalls should here be discussed, not least to clarify how the Cosmological Argument can proceed without its adherents succumbing to them. One such pitfall is discussed by William Rowe. Thus it would be fallacious to argue that since each material object has a causal explanation, so too must the collection of material objects[6] (and this fallacy forms a recognizable version of the reasoning Phillips objects to). This reasoning embodies the Fallacy of Composition, in which the qualities of members are fallaciously ascribed to the whole that they compose, as in the argument: The children in this school are small; therefore this is a small school. However, as Rowe remarks, adherents of the Cosmological Argument can avoid this fallacy, as did Samuel Clarke in the early eighteenth century, by arguing not to but from the set of material objects (and the fact that it could have had other members or even none at all), and applying to this a version of the Principle of Sufficient Reason (see Chapter Three, Section 3 above), without resort to the explicability of individual material objects.[7] If, as the Principle of Sufficient Reason implies, whatever could have been otherwise and is capable of explanation has an explanation (except where there is a sufficient reason for no explanation being possible), then this is true of the set of material

4 Terence Penelhum, 'Divine Necessity', p. 185.

5 Phillips, *Religion and the Hermeneutics of Contemplation*, p. 83.

6 William L. Rowe, 'Two Criticisms of the Cosmological Argument', p. 446.

7 Rowe, ibid., pp. 446–7.

objects being as it is, irrespective of the possibility that someone might reason fallaciously to this being the case.

The related charge is sometimes made against the Cosmological Argument that it moves illicitly from: 'Every material object has a causal explanation' to the conclusion: 'Therefore there is one causal explanation for all material objects'. Here there is what is known as a 'quantifier shift fallacy', from every object having some explanation or other to there being one particular explanation that all objects have, a fallacy exemplified in the argument that because everyone has parents, there is one pair of parents common to everyone. Theists would get nowhere if they argued along these lines. But in fact they have no need to disregard the scientific explanations of the existence of particular things, such as geological forces and evolution by natural selection. Consistently with accepting such explanations, they can raise cosmological issues by asking what brings it about that the set of material objects has the members that it has, rather than none at all; and asking questions of this kind involves a quantifier-shift fallacy no more than it involves a fallacy of composition.

But do such questions make sense? Does it make sense to ask for an explanation of collections such as the collection of material objects? This worry may underlie Phillips's reluctance to reason from the world as whole, and was earlier expressed by Bertrand Russell in his BBC debate with Father Copleston. When asked to explain how he could rule out 'the legitimacy of asking the question how the total, or anything at all, comes to be there', Russell replied: 'I can illustrate what seems to me to be your fallacy. Every man who exists has a mother, and it seems to me your argument is that therefore the human race must have a mother, but obviously the human race hasn't a mother – that's a different logical sphere.'[8] Russell may well be suggesting here that the very question that seeks an explanation of the collection of material objects fails to make sense, and involves what Gilbert Ryle later called a 'category mistake'. However, to take up Russell's example, although humanity has no mother, explanatory questions about the existence of human beings make good sense, and may well be answerable by reference to Darwinian explanations. There might have been no human beings, and these explanations help explain the fact that there are. The set of material objects is somewhat different from the set of human beings, partly because its members are not known all to be causally related to each other (although they may be, if the Big Bang theory is true). Nevertheless this set too could have had no members, and thus the question of why it has members apparently makes good sense, even though in this case the answer cannot be supplied by science. Thus Russell's scepticism about the meaning of such questions was misplaced. While some questions about some collections are meaningless (such as 'Does the human race have a mother?'), the relevant ones make good sense.

---

[8] Rowe, ibid., p. 445; Bertrand Russell and F.C. Copleston (1964), 'The Existence of God: A Debate between Bertrand Russell and Father F.C. Copleston', in John Hick (ed.), *The Existence of God*, 167–91, p. 175.

Rowe further discusses the objection that while collections can be explained, this can only be done by explaining the several members, and that no further explanatory question arises once this has been done. This objection will be considered in the coming section, involving as it does questions of explaining infinite sets. But before this matter can be considered, a misconception should be addressed. For it is often assumed that the Cosmological Argument assumes that the material universe had a finite past, and argues on this basis that to explain its origins appeal must be made to a first cause. Indeed the first three of Thomas Aquinas' Five Ways embody the assumption that an infinite series of causes or of other explanatory factors is impossible.[9] Thus a recent work which favours 'supernaturalism' (albeit not on the basis of cosmological reasoning) and discusses with aplomb the Fine-Tuning Argument (to be considered here in Section 5 below), nevertheless assumes that the Cosmological Argument rests on the 'intuition' that 'there cannot be an infinite regress of causes'.[10] But many proponents of the Cosmological Argument do not accept this 'intuition', let alone deploy it in their reasoning. Relatedly many proponents of this Argument do not assume that the material universe has a finite past, and, as was mentioned in Chapter One, Section 6 and Chapter Four, Section 4, this belief is in no way essential to theism. But without this assumption, the 'intuition' that an infinite regress of causes is impossible has no relevance to the argument whatever. Instead, the argument needs to cover both the possibility that the material universe is or has been finite and the possibility that it is infinite; in each case, it claims, the set of material objects stands in need of explanation.

One implication of this has already been discussed in Chapter Four. This concerns the kind of causality on which the argument pivots. If the Cosmological Argument is consistent with the material universe having had an infinite past, it cannot be claiming that the earliest physical event was caused to happen in much the same sense as that in which succeeding physical events were caused to happen by the earliest physical event. (Probably Aquinas was not suggesting this either. But if the Argument does not argue to an earliest cause at all, the implication becomes much clearer.) The kind of causal explanation to which the Argument reasons, as was argued in Chapter Four, Section 4, is rather a bringing about of what would otherwise not have been the case, and this kind of explanation is just as relevant to the existence of later as to that of earlier objects. The Argument concludes that all material objects are, have been or will be dependent on a creative agent with the power to make the difference between their existence and their non-existence, or to bring them about, and maintains that such bringing about supplies the needed explanation of the set of material objects. How the creator would bring this about the Argument does not claim to know; but that there is a creator with the necessary independence and power to do so is all that it claims and needs to claim.

---

9 Anthony Kenny (1969), *The Five Ways*, pp. 6–69.

10 Michael C. Rea (2002), *World Without Design*, p. 219.

Section 2: The Cosmological Argument

One premise of the Cosmological Argument, then, is that the set of material or physical objects has particular members, but could have had others, or none at all (the sheer 'gratuitousness of things', as it has been called).[11] Another premise is that it is possible that there is an explanation of this (a premise which once again seems hard to resist). For the other main premise we need to turn to the Principle of Sufficient Reason, which, according to Stephen T. Davis, was already implicit in the versions of this argument put forward by Aquinas in the thirteenth century, and became explicit in the versions put forward by Leibniz, Clarke and contemporaries of theirs in the eighteenth century.[12] The version of this Principle considered above in Chapter Three, Section 3 (and borrowed from Robert Nozick) runs as follows: there is a sufficient reason for whatever could have been otherwise, except where there is a sufficient reason to the contrary. Given that if an explanation is possible then there is no sufficient reason for there being no sufficient reason, there will have to be an explanation of the set of material objects (in the sense of the initial premise), and (for reasons that we have encountered in Chapter Four) nothing but the existence and agency of an immaterial, omnipotent or all-powerful, uncreatable individual will serve as such an explanation.

The premise likeliest to be contested here is the Principle of Sufficient Reason, which, as was recognized in Chapter Three, cannot be proved to be true, partly because any attempt to prove it is prone to assume its truth from the outset.[13] Indeed Richard Taylor argues persuasively that this Principle is a presupposition of all reasoning, or, as he puts things, 'of reason itself'; and this he says of a stronger version of the Principle (too strong and unqualified, or so I shall shortly claim), that there is some sort of explanation for everything. His argument is that this Principle must be a presupposition of reason, since one cannot argue for the Principle without already assuming it.[14] If we add that one cannot argue against it either without making the same assumption (qualified as in the coming two paragraphs), Taylor's conclusion seems even more secure. Davis, however, holds that the Principle is prone to be accepted only by those who accept the conclusion of the Cosmological Argument, and thus that the Argument may be regarded as begging the question. However, he is not suggesting that the Principle is derivable from this conclusion (rather than the other way round), and he recognizes that some atheists accept the Principle, although (as he also says) some (like Russell) do not.[15] Granted that the Principle is a presupposition of reasoning, the problem

---

11 Herbert McCabe (1992), 'The Logic of Mysticism – I', in Warner, *Religion and Philosophy*, 45–59, p. 52.

12 Stephen T. Davis (1997), *God, Reason and Theistic Proofs*, p. 69 and 77, n. 21.

13 Davis, p. 144; Richard Taylor [1964] (1974), *Metaphysics* (2nd edn), pp. 104–5.

14 Taylor, ibid., p. 105.

15 Davis, op. cit., pp. 145–6.

for those who reject the Principle is that if they reason at all, and thus presuppose the Principle, they are committed to accepting it, and if they also reject it they are inconsistent. Thus it seems reasonable to regard this Principle as given, even though it cannot be proved.

Some of the problems may be due to this Principle being expressed too strongly. Thus there is no reason to believe that what could not be otherwise can invariably be explained, although this would be implied by 'there is some sort of explanation for everything'. Besides, there are sometimes reasons why there cannot be an explanation. William Rowe has supplied the example of the fact that there are contingent states of affairs. For an explanation would have to be either contingent or necessary. But a contingent explanation would itself embody one of the contingent states of affairs to be explained, while no necessary state of affairs can explain contingent states.[16] Thus not everything that could have been otherwise has an explanation; in this case a sufficient reason can be given to the contrary. Another case is the existence of God, which, despite God's eternity, uncreatability and indestructibility, has to be regarded as logically contingent (see Chapter Four, Section 3 above). However, it makes no sense to ask for a reason for the existence of a being who is eternal, independent, uncreatable and indestructible. Once again, there is a sufficient reason for there being no sufficient reason. (Indeed this case would also be enough to establish what was just concluded about the previous case. It also blocks the objection to the Argument that a reason needs to be given for the existence of the creator.)

Accordingly the Principle of Sufficient Reason should not be expressed so strongly as to extend to whatever is the case, and should exclude cases where there is a sufficient reason for there being no sufficient reason. But with this qualification made, the Principle remains strong enough to cover sets or collections whose members are existing beings which can be caused to exist,[17] and maintains that in such cases there is always an explanation of the set or collection having members. And this is a reasonable claim; for there might have been nothing, and it is reasonable to seek an explanation for why there is instead the world that we find around us.

It is time now to return to an objection to the Argument first presented by David Hume. This is the objection that while collections can indeed be explained, this can only be done by explaining the several members, and that no further explanatory question arises once this has been done. As Hume expressed the point: 'Did I show you the particular causes of each individual in a collection of twenty particles of matter, I should think it very unreasonable, should you afterwards ask me, what was the cause of the whole twenty. This is sufficiently explained in

[16] William L. Rowe (1971), 'The Cosmological Argument', *Noûs*, **5**.1, 1971, 49–61, pp. 58–9.

[17] For a similar version of the Principle of Sufficient Reason, see Rowe, ibid., p. 59.

explaining the causes of the parts.'[18] So the question arises of whether a set or collection is sufficiently explained when its members have been explained. Finite and infinite sets or collections need to be considered separately.

Hume's example is drawn from a finite collection. Here a key point is that explanations are certain to involve reference to non-members of the set in question. Even if most members can be explained by another member of the set, this cannot apply to them all, or there would be 'a circle of causes', which is universally recognized to be impossible. So at least one member of the set must be explained by something external to the set. (Similarly, if the material universe were finite in extent or duration, its explanation would have to refer to something beyond itself.) Nonetheless Hume seems right to assert about sets of this kind that to explain each member is also to explain the set; for such explanations explain the set having these members rather than other members, and the set having some members rather than none.[19]

But as Rowe explains, matters are different for infinite sets. In this case, explanations of individual members of the set need never refer to non-members of the set, for there may always be some other member of the set (such as an earlier one) which serves to explain any given member. To adopt Rowe's example, even if there had always been human beings, every single human being could be explained as the offspring of earlier human beings and thus by other members of the set. But as Rowe adds, our discovering this would fail to explain why there are any human beings at all, rather than none. To try to answer this question by remarking that there had always been human beings would be to fail to answer the question on hand (that of why there are any human beings at all). Hence Hume's suggestion about what is involved in explaining collections, while apparently true of finite sets, fails when it comes to infinite sets.[20] Indeed in both cases there has to be something outside the set which explains the set having the members that it does rather than none.

This rather complicated reasoning turns out to be directly relevant to explaining the material universe. If it has had a finite past, at least its earliest members or states would stand in need of explanation from something beyond the temporal sequence. But even if it has an unlimited past, the fact that it has members at all these times (rather than none) is equally in need of explanation, and no adequate explanation is to be found by reference to other material objects, or thus by reference to factors (such as evolution and natural selection) discoverable by the methods of empirical science. Granted the Principle of Sufficient Reason, there has to be an explanation of the material world which lies beyond natural agents and agencies. Nothing less than a Creator will do.

---

18 David Hume, 'Dialogues', Section IX, Wollheim, op. cit., p. 164. The matter is further discussed in Robin Attfield (2004), 'Rousseau, Clarke, Butler and Deism', *British Journal for the History of Philosophy*, **12**.3, 429–43.

19 Rowe, 'Two Criticisms', p. 246.

20 Rowe, ibid., 256–8.

Hume presents some further objections to the Cosmological Argument. But several of these turn on this being an argument to a logically necessary being (as it is in the hands of Clarke and of Leibniz), and the objection that there are no such beings.[21] This objection, however, is beside the point as long as God's existence is not understood as logically necessary (see Chapter Four, Section 3 above), and thus makes no impact on versions of the argument that avoid this concept, for example, those of Aquinas and of Clarke's contemporary, Daniel Waterland,[22] or later that of Jean-Jacques Rousseau.[23] If adherents of the argument were nevertheless to introduce a logically or conceptually necessary being, the objection would still carry weight; but in fact the introduction of such a being actually conflicts with the aim of the argument to produce an explanation for certain contingencies, since no necessary state of affairs, as we have seen, can explain contingent states of affairs. Thus it was ill-advised on Hume's part to make his character Cleanthes 'rest the whole controversy' on this issue.[24]

Hume's Cleanthes additionally raises objections to *a priori* arguments being employed to establish matters of fact, an objection echoed recently by Dupré.[25] But it is unclear that the Cosmological Argument has a significantly *a priori* character. Its premise concerning the set of material objects having the members that it does is contingent; again, given that its conclusion avoids the concept of a logically necessary being, it also avoids the apriorism of the Ontological Argument. And if the Principle of Sufficient Reason is held to be *a priori*, that no more vitiates this argument than it vitiates other arguments about states of affairs that employ one or another explanatory principle of this kind, such as the original argument for the existence of the planet Pluto (discovered on this basis before that planet was observed), or the argument for there having been a continent (usually called 'Pangaea') that preceded continental drift. Besides, since we all presuppose this Principle, there can be nothing objectionable in resorting to it, whether what is at issue is a matter of fact or of logic.

The Cosmological Argument, then, is unaffected by Hume's objections. It is not a proof, since one of its premises, the Principle of Sufficient Reason, is unproven; but it supplies strong grounds for its conclusion. It does not supply explicit grounds for belief in God having a will or purposes or personality; in these regards, theistic belief depends on other foundations. Nevertheless it supplies strong grounds for the existence of an all-powerful, all-knowing, independent, uncreatable and indestructible agent whose agency brings about the material

---

21 Hume, op, cit., pp, 162–3.

22 For Aquinas, see Patterson Brown, op. cit.; for Waterland, see Robin Attfield (1993), 'Clarke, Independence and Necessity', *British Journal for the History of Philosophy*, **1**.2, 67–82, pp. 68–73.

23 Jean-Jacques Rousseau [1762] (1963), *Emile*; Robin Attfield, 'Rousseau, Clarke, Butler and Deism' (see note 18 above).

24 Hume, op. cit., pp. 162–3.

25 Hume, ibid., p. 162; Dupré, op. cit., p. 44.

world and sustains it in being,[26] and for a being beyond experience that makes the experience on the part of creatures possible, indeed for an agent most readily understood as personal. As such, it supplies grounds for central aspects of belief in creation.

Section 3: Varieties of Design Argument

The Design Argument (or Teleological Argument) is the only argument for the existence of God given serious consideration by Dupré, since it resembles scientific arguments in arguing from particular observable phenomena to what is considered their best explanation. According to Dupré, it is when we recognize that the same phenomena are better explained by reference to natural processes, as by Darwin, that we become free to discard belief in a supernatural designer as groundless; indeed Dawkins has declared that Darwinism at last makes it possible to be 'an intellectually fulfilled atheist'.[27] Dupré writes of 'the Argument from Design', but I will avoid this phrase, since on some accounts talk of design presupposes talk of a designer, and thus begs the question. What is intended here is actually adaptation, the suitability of organs, organisms or species for their biological role or function, where 'function' can be neutrally specified without reference to the purposes of a designer. It was precisely such adaptation that Darwin's theory of natural selection was introduced to explain.

While design arguments based on what would now usually be called adaptation have often been advanced, particularly in the eighteenth and nineteenth centuries, it should not be assumed that the design argument need have such a basis or structure. For there are equally longstanding arguments to design from the phenomenon of order across space and time in the form of the universal applicability of laws of nature, and there are more recent arguments from the apparent fine tuning of universal variables to the conditions necessary for life. Besides, not everyone considers that arguments concerning design must proceed by analogy from recognized cases of design to the world as a parallel case. Nevertheless examples should first be given of arguments of just such an analogical character, with the phenomena of adaptation as their starting-point.

Hume's character Cleanthes, for all his rejection of the Cosmological Argument, presents such a version of the Design Argument, arguing from cases of apparent design in the manner of a succession of Boyle Lecturers of earlier in the eighteenth century, and appealing to common sense: 'Consider, anatomize the eye; survey its structure and contrivance; and tell me, from your own feeling, if the idea of a contriver does not immediately flow in upon you with a force like that of

---

[26] See further Attfield, *God and The Secular*, pp. 159–81.

[27] Dupré, op. cit., pp. 47–55; Richard Dawkins (1986), *The Blind Watchmaker*, p. 6.

sensation.'[28] The eye is here held to resemble a device engineered for a purpose; the analogy is so striking that a purposive explanation of such adaptation has to be accepted. (Such an argument by analogy is, of course, quite distinct from the theory of analogy, the analogical account of the meaning of predicates used of God, presented above in Chapter One, Section 6.)

Despite Hume's devastating criticisms of such an argument (see Section 4 below), it was classically presented some decades later by William Paley in a form that greatly impressed Darwin in his youth and influenced the work of his maturity. Paley's version is the sole form of the Argument considered by Dupré:

> In crossing a heath, suppose I pitched my foot against a stone and were asked how the stone came to be there: I might possibly answer that for all I knew to the contrary it had lain there forever: nor would it, perhaps, be very easy to show the absurdity of this answer. But suppose I had found a watch upon the ground, and it should be inquired how the watch happened to be in that place; I should hardly think of the answer which I had before given, that for any thing I knew, the watch might have always been there.[29]

Having quoted this passage, Dupré proceeds to paraphrase Paley as follows:

> I should, of course, conclude that someone had designed and constructed the watch. Paley then points out that elaborate though the workings of a watch may be, they are simple compared to those of a plant or an animal. Consequently the argument for a designer in the latter case is stronger than that in the former. Hence nature must have been designed and created, and this designer and creator we refer to as God.[30]

Once again, the analogical character of the argument is clear. Strictly, the argument supports, if anything, belief in a designer or 'artificer', as Paley recognized, rather than in a creator; but if there are independent grounds from the Cosmological Argument for belief in God as an all-powerful, uncreatable and indestructible agent, then the conclusion of the Design Argument could readily be regarded as supplementing the conclusion of that argument (with which it is fully compatible).

However, a different design argument, not based on adaptation, had also long since been advanced by Isaac Newton and his followers. This was an argument from the motions of the heavenly bodies, which suggest a designer with a knowledge of mechanics.[31] Newton's argument incorporated other phenomena, such as the solar system as a whole: 'This most beautiful system of the sun,

---

[28] Hume, 'Dialogues', Section III, Wollheim, p. 28; see also Section IV, Wollheim, p. 137.

[29] William Paley (1825), *Natural Theology*, in Paley, *The Complete Works of William Paley, D.D.*, vol. III., pp. 3–4.

[30] Dupré, op. cit., p. 48.

[31] Robert H. Hurlbutt (1965), *Hume, Newton and the Design Argument*, p. 13.

planets, and comets, could only proceed from the counsel and dominion of an intelligent and powerful being';[32] comparable arguments from the recent period will later be encountered. Newton also argued from biological adaptations such as the eye[33] But as the discoverer of laws of motion probably his most original contribution to the design argument lay in reasoning from the order displayed in such 'mechanical principles' to the laws created by a designing deity. Hurlbutt regards Hume's criticism of the design argument as directed at the Newtonian version as well as at the older, biological version.

Yet the medieval version of the Design Argument, found in Thomas Aquinas, is sometimes regarded as parallel to Newton's, although unable to appeal to laws of motion. Interpreting Aquinas' claim in his Fifth Way for upholding the existence of God that 'an orderedness of actions to an end is observed in all bodies observing natural laws', Keith Ward rejects as implausible the interpretation that all events are observed to have final causes (or purposes). He adopts instead the view that what Aquinas has in mind is simply the orderedness of physical law, and the fact that objects obey such laws rather than moving at random.[34] Thus the argument from the order of natural laws may go back to Aquinas. In any case, it cannot be ignored in appraisals of the Design Argument. To the extent that this is an analogical argument, the analogy is not with a machine but with the order, spread out across time, of a dance, song or symphony.

Ward further suggests that the Design Argument should not be treated as an argument from analogy, bearing what he considers the weaknesses of such arguments (such as extrapolation from selected aspects of the universe to the universe as a whole). Instead, the argument should be treated as an exercise in rational interpretation, in which defensible concepts such as explanatory intelligibility are applied to reality with a view to discovering the best form of explanation;[35] and this certainly introduces a distinctive approach to design arguments. To carry through this programme, Ward investigates the valuable end-states that a designer might aim to embody in the world; here, however, consideration of this argument from value is deferred to a later chapter. But as Hurlbutt says, there is no need to appeal to desired end-states to consider arguments for design, for example, from the order displayed in laws of nature.[36]

More recent versions of the Design Argument should also be mentioned. One is that of Michael Behe, mentioned in Chapter Four, Section 5. Behe argues from biological cases of (what he calls) irreducible complexity, maintaining that the gradualism of the theory of evolution by natural selection cannot always

---

32 Newton, *General Scholium* of the *Principia*, quoted at Hurlbutt, p. 14.

33 Hurlbutt, pp. 12–13.

34 Aquinas, *Summa Theologiae*, qu. 2, art. 3; quoted in Keith Ward, *Rational Theology*, p. 100.

35 Ward, ibid., pp. 94–6.

36 Hurlbutt, op. cit., p. 16.

account for the kind of complexity where in the absence of any single factor a whole system would be impossible (an argument comparable in some ways to Newton's argument from the beauty, grandeur and complexity of the solar system).[37] But this argument, rather like that from adaptations, is open to the kind of trenchant criticism levelled by Michael Ruse, for its author's inadequate grasp of relevant tracts of science and for prematurely abandoning naturalistic explanations in favour of supernatural ones. For example, the Krebs cycle, by which food is converted into usable energy, supposedly an example of irreducible complexity, can in fact be explained as evolving through a series of adaptive stages,[38] as also can Behe's favourite example, the blood-clotting system.[39] Besides, if cells were long ago specially created with powers suited to complex processes before creatures needing these processes originated, as Behe supposes, it is hard to explain how these powers survived and were not eliminated by processes such as natural selection or genetic drift. Thus at least some of the arguments for 'Intelligent Design' are unimpressive.[40]

Twentieth-century arguments also include the argument from Fine Tuning and from the conditions necessary for life to originate and survive.[41] Reasoning of this kind could be construed either as an argument about particular biological systems (for example, biochemical ones), and their systemic adaptedness to life (in which case it might be open to the criticism just mentioned), or as an argument concerning laws of nature, this time appraised by their conduciveness to the emergence of life and its survival. In its latter form, the argument need not turn on adaptation or involve an appeal to supernatural interventions, and would rather concern the general character of nature, understood in the argument's conclusion as a habitat designed for life. (Despite similarities, this argument should not be confused with the 'anthropic principle', which is directly concerned with the conditions necessary for the existence of human life only.[42]) Michael C. Rea remarks that this argument turns on far more than the improbability of the outcome, and that a detailed assessment of the various probabilities is yet to

---

[37] Michael Behe, op. cit.

[38] Kenneth Miller (2004), 'The Flagellum Unspun: The Collapse of Irreducible Complexity', in William A. Dembski and Michael Ruse (eds), *Debating Design*, 81–97.

[39] Thus Kenneth Miller in Miller (1999), *Finding Darwin's God*; cited in John Hedley Brooke, 'Revisiting Darwin, p. 48.

[40] Ruse, *Can a Darwinian*, pp. 115–22.

[41] Davis, *God, Reason and the Theistic Proofs*, pp. 107–15; John Leslie (1989), *Universes*.

[42] John D. Barrow and Frank J. Tipler (1986), *The Anthropic Cosmological Principle*, p. 15. In its 'weak' form, the Anthropic Principle is sometimes used to suggest that there is nothing surprising about these conditions being satisfied: see Le Poidevin, *Arguing for Atheism*, pp. 54–6. To this suggestion, Stephen Davis produces formidable replies: see Davis, op. cit., pp. 112–13. (We return to the Fine Tuning Argument in the coming section.)

be produced.[43] Nevertheless these recent versions, as well as those of Cleanthes and Paley, Newton and Aquinas, need to be remembered when criticisms of the Argument from Hume and from modern Darwinians are assessed (see the coming two sections and the two chapters following).

The variety of versions of the argument is important. While Dupré mentions one version only (Paley's), another recent critic of design arguments, Robin Le Poidevin, confines himself to the argument from adaptations (that of Hume's Cleanthes and of Paley) and the Fine Tuning Argument, plus versions of the anthropic principle, as opposed to the argument from natural regularities as a manifestation of order.[44] Yet this latter argument too deserves attention (and receives some in the coming section). People can hardly form their own view of design arguments in general if they learn only of the more historically dramatic versions (Paley's) or of those that are topical (arguments, say, from 'irreducible complexity' and from Fine Tuning, the ones given prominence in an excellent recent number of the journal *Think*).[45] Hence the survey and review of versions of the current section.

Section 4: Hume's Critique of the Design Argument

Some of Hume's criticisms relate to principles of explanation. His character Philo suggests that we cannot seek to explain what is unique.[46] But, as Swinburne has replied, this is a mistake, for otherwise we could not seek to explain the origins of the human race, simply because of its apparent uniqueness.[47] The corresponding suggestion that nothing unique or undiscovered can be invoked as an explanation[48] also belies scientific practice. When distinctive phenomena are to be explained, distinctive explanations need to be invoked,[49] as when the planet Pluto was first discovered in this way.

Hume's view that explanations must be proportioned to the phenomena to be explained[50] is more acceptable. But it does not imply, as Philo suggests, that an embodied designer would suffice.[51] Even if some localized adaptations (such

---

43 Rea, op. cit., pp. 215–18.

44 Le Poidevin, *Arguing for Atheism*, pp. 44–58.

45 The Autumn 2005 number of *Think: Philosophy for Everyone* (number **11**) carries papers on 'irreducible complexity' and on the Fine Tuning Argument from several leading thinkers.

46 David Hume [1777], *Enquiry concerning Human Understanding*, Section XI; in Hume (1975), *Enquiries*, ed. L.A. Selby-Bigge (3rd edn), p. 148.

47 R.G. Swinburne (1968), 'The Argument from Design', *Philosophy*, **43**, July, 199–212, pp. 207–8.

48 Hume, op. cit., Section XI, Selby-Bigge, p. 148; see Swinburne, op. cit., pp. 207–8.

49 Swinburne, op. cit., pp. 207–8.

50 Hume, *Enquiry*, Section XI, Selby-Bigge, p. 136.

51 Hume, Dialogue V, Wollheim, p. 142.

as the eye) could in theory be explained in this way, laws of nature concerning gravity, light, electricity and sound and applying across space and time could not. Besides, as Peter Payne has remarked, 'Physically embodied designers would seem to be dependent upon the fundamental laws of physics and hence could hardly be invoked to explain them'.[52]

Other Humean criticisms concern whether order requires to be explained. Given long enough, argues Hume's Philo, order can be generated by random processes, and does not need a purposive explanation.[53] While this claim may be true of adaptations, and could even be held to anticipate Darwin, it implicitly relies, as Swinburne has observed, on the operation of natural laws; but such laws cannot themselves be explained in this way, for it would be a category mistake to suggest that they are capable of evolving. Rather they apply across time, as, for example, astrophysicists assume. Hence the kind of order embodied in laws of nature cannot be explained by random processes (nor, come to that, by non-random ones either).[54] Hume's suggestion that the natural order may be self-sustaining[55] and implicitly self-explanatory is subject to the same comments. We should not abandon the search for explanations, except where it is an impossible one.

Similarly Hume's reflection that the world's order resembles that of an organism as much as that of a mechanism, and that there is no need to invoke a designer to explain the former,[56] may be relevant to adaptations (particularly as the order present in organisms is nothing but biological adaptation generalized); but it again relies on the operation of natural laws, which do not evolve, and which cannot be explained in this manner. Hume's even more Darwinian remark that in the absence of apparent design animals could not have survived, and that their adaptedness is therefore unsurprising,[57] may help explain adaptations, but implicitly concedes that the world's processes are regular enough for such adaptations to have emerged; as Davis comments, the world could easily have been much less regular than it is.[58] The same comments apply to Hume's further suggestion that any universe must seem ordered, and to Kant's view that order is generated by the human mind.[59] For it is not difficult to describe possible worlds (and ones like ours at that, incorporating components classifiable by us) not complying with laws of nature; we could not survive in them, but that does

---

52 Peter Payne (1996), *Design in the Universe*, cited in Davis, op. cit., p. 103.

53 Hume, Dialogue VIII, Wollheim, p. 156.

54 Swinburne, op. cit., p. 202.

55 Hume, Dialogue VIII, Wollheim, p. 156.

56 Hume, Dialogue VII, Wollheim, pp. 152–4.

57 Hume, Dialogue VIII, Wollheim, pp. 156–58; see also Davis, op. cit., p. 102.

58 Davis, ibid., p. 102.

59 Hume, Dialogue VIII, Wollheim, p. 156; Immanuel Kant, *Critique of Pure Reason*, op. cit. (1968), pp. 65–119; see also Thomas McPherson (1972), *The Argument from Design*, pp. 17–20.

not remove the need to explain how it is that the actual world differs from these possible ones.[60]

Hume further objects that if the world's order and complexity requires a designer, then so does that of the designer.[61] This seems implicit either in the Principle of Sufficient Reason or in any principle of explanation which requires an explanation for cases of order, as the design argument undeniably does. (Yet, as Michael Ruse remarks in much the same context, we need not accept the related claim of Richard Dawkins that complexity needs greater complexity to explain it: see Chapter Seven, Section 4 above.[62]) Now if the designer is God, then as we have already seen, it makes no sense to ask for an explanation of the existence of an independent, uncreatable and indestructible being. Certainly, if the design argument is considered alone, then the designer could in principle have an explanation (such as being made by God), despite the complexity of supposing this to be the case. Yet as Davis points out, this would not weaken an explanation that invokes a designer.[63] Even if, as in the case of human designers, an ulterior explanation of this designer's existence can be requested, explanatory weight is still supplied when a puzzling phenomenon is attributed to contrivance, conspiracy or planning. In this case, however, the design argument need not be considered in isolation; and if the designer of this argument is identical with the deity of the Cosmological Argument, then the request for a further explanation turns out to be nonsensical after all.

Even so, Hume's further criticism of the argument is well taken that the argument does not prove the God of theism. Even if we set aside the issue of embodiment, the designer is not known on the strength of the argument to be infinite, omnipotent, omniscient, or eternal.[64] While this should be granted, the success of the argument in establishing the existence of a designer of great power and intelligence would still raise the issue of whether these further attributes apply to that designer, an issue of which the solution would then depend on other available grounds.[65] Hume also suggests that the world's order might be explained not by one designer but by a consortium.[66] This is not impossible, but, as Swinburne argues, it would only become credible if there were relevant evidence, such as one tract of space or time complying with one set of laws and other tracts with different ones; in the absence of such evidence, this is an absurdly uneconomical hypothesis.[67] Indeed insofar as the phenomenon to be explained is the uniformity of the laws of nature observed at all times and places,

---

60 Attfield, *God and The Secular*, p. 186.

61 Hume, Dialogue IV, Wollheim, pp. 134–6.

62 Dawkins, *The Blind Watchmaker*, p. 141; Ruse, *Can A Darwinian*, p. 114.

63 Davis, op. cit., p. 101.

64 Hume, Dialogue V, Wollheim, pp. 139–40, 142.

65 Davis, op. cit., p. 103.

66 Hume, Dialogue V, Wollheim, pp. 140–41.

67 Swinburne, op. cit., p. 210.

the hypothesis of a committee of designers, while remaining possible, can hardly be regarded as the best explanation.

Hume's Philo further argues that the existence of evil and suffering counts against the design argument, as it casts doubt on the designer's goodness.[68] As was seen in Chapter Four and will be seen again in the next chapter, considerations of this kind proved significant for Darwin. Certainly most versions of the design argument contribute little to solving the problem of evil. Ward's argument from value to divine purpose may be different, but is yet to be considered. (It figures in Chapter Eight.) However, the design argument itself is unaffected by the fact of evil, for its conclusion says nothing about the moral character that a designer would have. The designer would have to be a lover of order, but need not love humanity, nor (except where the argument from Fine Tuning, which was unavailable in Hume's day, is concerned) life itself. Thus if the argument succeeds, questions about the designer's character remain to be addressed. The issues raised by the problem of evil are important ones, but are here deferred pending consideration, together with a cluster of issues about disvalues in nature, in Chapter Six; these issues include that of the sense of 'goodness' as applied to God.

The final objection of Hume to be considered is possibly the most important, the objection that the analogy is too weak to support its conclusion. The similarities between the universe and a machine are matched by dissimilarities, and stressing the former is no more reliable than stressing the latter.[69] This is part of the point of previous criticisms. Thus if we had other universes with which to compare the actual one, we might be better placed to analogize to design as a conclusion, but in the nature of the case we only have one. If order invariably betokened design, or design bespoke a single designer, or the designer were more like familiar embodied human designers, or if (say) the explanation more clearly fitted the phenomena one for one, the argument would be more impressive. As things are, design by an intelligence seems just one of the hypotheses compatible with the phenomena. These previous criticisms, however, have been found to be inconclusive, except for the new one about explanations fitting the phenomena one for one; and this degree of fit is not demanded for purposive explanations in human social life (in police investigations of crimes, for example).

Certainly the similarities between the universe and a watch or other mechanical device have their limits. But Hume's other objections already overthrow the argument from biological adaptations, and thus versions of the argument from spatial structures. Matters are different with versions of the argument concerning the order displayed in laws of nature, order compared above not with a machine but with the order present in a dance, a song or a symphony. They are also different for versions concerning the apparent fine tuning of such laws to the conditions necessary for life (see the next section), where the comparison is with an intelligent

---

[68] Hume, Dialogues X and XI, Wollheim, pp. 165–88.

[69] Hume, Dialogues IV and VIII, Wollheim, pp. 134 and 160.

listener tuning a radio or an intelligent mechanic tuning an automobile engine. These versions of the argument present much stronger analogies to design on the part of a single intelligence than is to be had from the version from biological adaptations; indeed the fine tuning version has been re-expressed, as William Lane Craig does, in terms of probabilities:

> The point is that it is unimaginably more probable that the universe should be life-prohibiting than life-permitting, and that the best explanation of the cosmos as it is may well be intelligent design.[70]

For if certain cosmic variables were different from their current values by a millionth either way, life would be impossible.[71] (The sense of 'probable' here is discussed in the section that follows.)

The argument can still be construed as an argument by analogy. If so, while Hume's objection to its weakness cannot be regarded as conclusive, the argument continues to harbour the weakness of comparing parts of the universe with the universe as a whole; this is not irrational, particularly in matters of explanation, but invites critics to target the apparent lacuna implicit in such reasoning. However, there is an alternative possibility, implicit in Craig's words (just quoted) and also in Ward's interpretation. The argument can be understood as having two stages, a first stage, adverting to the laws of nature, appealing either to the Principle of Sufficient Reason or to any weaker explanatory principle that requires this order to be explained, and concluding that there must be some explanation of the remarkable fact of the world's order, and a second stage, of which the conclusion is that design by an intelligence is the best explanation. Thus construed, the argument by-passes Hume's objection concerning the weakness of Cleanthes' analogy, and, without purporting to enjoy the status of a scientific argument, complies with a model (argument to the best explanation) widely recognized in the philosophy of science. In this version the conceptions of a mind are construed as the best explanation of the cosmic scope of the laws of nature and their facilitation of life on Earth (and, for all we know, elsewhere).

Accordingly, none of Hume's objections prove fatal, or even significantly detrimental, to the design argument in general, although several of them undermine versions of that argument from biological adaptations (including those supposed to be 'irreducibly complex'). While Hume was in no position to respond to the Fine Tuning Argument, his critique makes little or no impact on the version of the design argument from laws of nature, which had been deployed since at least the time of Newton. Whereas biological adaptations can be explained by natural processes (including natural selection) operating over long

---

70 William Lane Craig (1993), 'A Criticism of the Cosmological Argument for God's Non-Existence', in Willliam Lane Craig and Quentin Smith, *Theism, Atheism and Big Bang Cosmology*, 256–76, p. 268.

71 Davis, op. cit., pp. 108–11.

periods of time, those processes themselves display a regularity unamenable to any explanation of this kind. And even if some of them can be explained by others, the fact that there are such regularities remains beyond naturalistic explanation, and yet comprises a striking fact, needing to be explained (whatever is held about their life-permitting character). So at least this version of the design argument survives Hume's criticisms.

Section 5: Fine Tuning, Universes and Probability

The argument concerning the apparent fine tuning of the laws of nature and of the fundamental constants of the universe to the conditions necessary for life warrants a further brief discussion. We thus return to Craig's remark that 'it is unimaginably more probable that the universe should be life-prohibiting than life-permitting', and that therefore design by an intelligence is the best explanation of the actual world.[72] What sense can be made of 'probable' in this passage, and how does this affect the argument?

A critic of this argument, Martin Rees, maintains that the life-permitting character of the actual world becomes much less surprising if we envisage our universe as one of many. (A universe would have to be regarded not as comprising all physical objects and processes, as hitherto, but instead all those objects and processes that result from a given Big Bang. Big Bangs that result in worlds causally unrelated to our own could then be regarded as generating other universes.) If universes are numerous, it becomes unsurprising that at least one of them is life-permitting. Given the view that probability consists in the relative frequency of a phenomenon in a population, the phenomenon of life in our universe also becomes no less probable than might be expected.[73]

To this criticism, there are several replies. The common sense reply is that of Stephen Davis: there is no evidence whatever that there are any other universes; indeed science relies on the laws of our universe applying across the whole of space and time (rather than there being other laws in some other region or at some other period).[74] Rees remains free to talk of *possible* (life-prohibiting) universes, but to make this move would be to reinstate the original argument, which says that the life-permitting character of the actual universe could easily have been lacking (and indeed is lacking in most possible universes), and needs an explanation. Secondly, probability as relative-frequency cannot be the relevant kind of probability. It is not just that we cannot finish counting possible universes, nor even that we cannot begin; for talk of 'unimaginably more probable' makes no reference to statistics in the first place.

---

[72] Craig, op. cit., p. 268.

[73] Martin Rees (2003), 'Other Universes: a Scientific Perspective', in Neil A. Manson (ed.), *God and Design*, 211–20.

[74] Davis, op. cit., p. 115.

Thirdly, it has been replied by D.H. Mellor that what is the case in other worlds, even if they were to exist, makes no difference to the intrinsic probability or improbability of the laws of our own, for the relevant kind of probability is physical probability. But the physical probabilities of events are moulded by laws of nature, and there can be no such probabilities when laws of nature cannot be taken for granted, as in Rees's response or in Craig's original argument. This not only means that Rees's response fails, but also that Craig's argument begins to seem incoherent; for there is no relevant concept of probability in which it could be expressed.[75] Le Poidevin presents a parallel argument: while talk of the probability of a life-permitting universe makes good sense on the hypothesis of divine design, it cannot be said to raise the probability of such a universe from what it would be given God's non-existence, since on the latter hypothesis a life-permitting universe is neither probable nor improbable. This is because there would be nothing either having or lacking propensities to generate a universe of any kind.[76]

But this is not the end of the matter. If the Principle of Sufficient Reason (adhered to by some religious sceptics like Lucretius as well as by many religious believers) is granted, then if there is also no creator we really can assign a probability to a life-permitting universe, that of nil, since the probability of any kind of physical universe would also be nil. (That this is so was explained in Chapter Three, Section 3.) This being so, the hypothesis of divine design really does raise the probability of a life-permitting universe. Here it could be objected that the concept of probability just employed cannot be regarded as physical probability, as it appeals beyond the laws of nature of the actual universe. But if any appeal beyond such laws is to be disqualified, and all associated talk of probability ruled out of order on the same ground, then the very question of whether the universe can be explained has been begged, together with that of which explanation, if any, is preferable, and the Principle of Sufficient Reason has implicitly (and prematurely) been ruled to be inadmissible at the same time. Since this is a metaphysical Principle, the concept of probability in use here could reasonably be regarded as metaphysical (rather than physical) probability, since it concerns probabilities that may or may not be grounded in the physical laws of the actual universe. This concept, together with antonyms (such as 'improbable'), synonyms (such as 'likely') and related terms (such as 'probably') will continue to be used as appropriate both below and in the coming chapters.

Even those who consider the Principle of Sufficient Reason as less than secure knowledge may nevertheless be prepared to accept some less demanding explanatory principle according to which the cosmic constants and the laws of nature of the actual universe are explicable, alongside remarkable phenomena in general, and thus that the existence of a life-permitting universe is (in a corresponding sense) improbable unless there is an explanation of why those

---

[75] D.H. Mellor (2005), 'Accepting the Universe', *Think*, **11**, 55–64. See also Le Poidevin, op. cit., p. 56.

[76] Le Poidevin, op. cit., pp. 53–4.

constants and laws are as they are. In other words, a life-permitting universe without a creator could on this basis be held to be metaphysically improbable, even without the Principle of Sufficient Reason itself being granted unreservedly. For there is in any case nothing incoherent about possible explanations of the cosmic constants and the related life-permitting character of the actual world (unless, with Mellor, we insist that explanations invariably turn on earlier events[77]). More than one explanation seems possible (design by a single intelligence *or* design by a consortium, just to cite examples mentioned above), and some kind of explanation seems to be needed. Indeed Le Poidevin remarks at this stage that even in cases where no physical probabilities are relevant (because laws of nature can no longer be appealed to), it may remain 'right to request a further explanation of fundamental constants', as long as no appeal is made to 'probabilistic considerations'.[78] But even if talk of probability were restricted to physical probability, and talk of metaphysical probability were thus not to be countenanced, at least until further study of the concept of probability (as advocated by Rea[79]) has been undertaken, the two explanations just mentioned could still be compared for cogency or credibility, and there could be reason to accept the more credible.

Thus even qualified sceptics about the Principle of Sufficient Reason may well credit that remarkably counter-intuitive states of affairs, such as the life-permitting character of the actual universe, require explanation, even if more humdrum contingencies somehow (contrary to the position defended above in Chapter Three) do not. And granted once that there *is* an explanation, the second stage of the argument (see the previous section) readily brings us to a single designing intelligence as the *best* explanation.

Accordingly, those versions of the design argument which turn on laws of nature rather than particular adaptations or systems confer genuine rational support upon belief in a designing intelligence, and thus supplement the conclusion of the cosmological argument (see Sections 1 and 2 above). Nevertheless, the problems for theistic belief generated by Darwinism and the issues surrounding disvalues in nature remain to be discussed. These issues are considered in the two chapters that follow.

---

77 Mellor, ibid., p. 60.

78 Le Poidevin, op. cit., p. 56.

79 See Section 3 (above).

Chapter Six

Darwinism, Disvalues and Design

Section 1: Darwinism

It is time to say more about Darwinism, partly to explain its coherence and cogency, partly to introduce its implications for the Teleological Argument, and partly to illustrate and appraise the problem of natural evil which, as we have seen, contributed to Darwin's own loss of belief in God. Darwinism will be discussed in the current section, together with some of its grounds and some of its varieties, to which we revert in later chapters. Its possible implications in the area of sociobiology will also be considered. In later sections of this chapter its bearing will be considered on the Design Argument, as will subsequently its compatibility with belief in God's goodness. The theory of evolution by natural selection provides the backdrop for contemporary environmental concern, and several of the themes of this chapter will thus prove relevant to the scope and limits of environmental stewardship, discussed in the final two chapters of this book.

Darwin's theory asserts, in part, that life on Earth has evolved from a common ancestry[1] and from relatively simple to relatively complex forms of life.[2] Michael Ruse summarizes Darwin's evidence as including grounds from palaeontology (the fossil record), bio-geography (for evolution best explains the distribution of bird and reptile species in places such as the Galapagos Islands), anatomical similarities across diverse species (for evolution from common ancestors best explains such 'homologies'), and from systematics, for common ancestry best explains the way species can be ordered in a hierarchy of groups, which seem to call for an explanation of such a nature.[3] In view of this and additional evidence (like that from embryology), the evolution of species was soon recognized as a fact.[4]

Darwin further argued that a major cause of the adaptedness of organisms that facilitates their evolution was natural selection. Because more individual organisms are produced than can survive, there is competition for survival, and only the fittest survive and reproduce. Darwin was hereby generalizing the findings of Thomas Robert Malthus to all biological populations, and deducing relevant implications. His central grounds, however, were an argument by analogy with

---

1 Michael Ruse (2001), *Can a Darwinian Be a Christian?*, pp. 12–18.

2 Alvin Plantinga (1991), 'When Faith and Reason Clash: Evolution and the Bible', p. 684.

3 Ruse, op. cit., pp. 13–18.

4 Ruse, ibid., pp. 20 and 25.

the artificial selection among animal and plant varieties conducted by human breeders who select and breed from the best available variants;[5] thus Darwin's analogical argument was to supersede the previously influential analogical argument of William Paley who had explained adaptations by a cosmic watch-maker or designer. More recently, Richard Dawkins has developed this analogical argument as supporting apparent design by a 'blind watch-maker', or in other words natural selection, adding to this conclusion a rejection of any other designer;[6] the suggestion (detected by some in the text of Darwin[7]) that natural selection may be regarded as personal and intentional is firmly excluded. The source of the variations on which natural selection operates was unknown to Darwin, who also adhered to mistaken beliefs about heredity; but when the work of his contemporary, Gregor Mendel, on genetics and on random mutations, was rediscovered at the turn of the century and developed by researchers such as Ronald Fisher in the 1930s, it became possible to explain both the origination of beneficial characteristics and their survival down the generations, and thus establish a broad Darwinian consensus among most scientists.[8]

However, critics of Darwinism sometimes complain that natural selection explains nothing, since the survival of the fittest simply means the survival of those fittest to survive, and this principle is an empty, unfalsifiable tautology. Yet, as Gallagher has replied, the theory of evolution (as opposed to this principle) supplies an informative explanation both of variations, of some traits being more advantageous than others, of selection between them, and of the related development of populations and species. Thus while Darwinists from Ruse to Stephen Jay Gould have tried but failed to make the actual principle into a substantive claim, the theory remains a substantive one. If this were not so, it could not be contested by its creationist and other opponents as misguided.[9] So the theory of natural selection cannot be written off as lacking any scientific role. Meanwhile the principle effectively tells us to seek explanations of biological survival in the area of an organism's adaptation to its natural environment; this is a methodological directive, rather than a statement, but one grounded in the theory of natural selection,[10] and thus not lightly to be sneezed at.

Darwin and his mainstream successors have taken the view that the variations that are important for natural selection are invariably small ones, despite the availability of larger variations, or 'hopeful monsters' as he called them; evolution

---

5 Ruse, ibid., pp. 22–3.

6 Richard Dawkins (1986), *The Blind Watchmaker*.

7 Gillian Beer (1983), *Darwin's Plots*, pp. 68–71.

8 Ruse, op. cit., pp. 25–6.

9 Kenneth T. Gallagher (1989), '"Natural Selection": A Tautology?,' *International Philosophical Quarterly*, **29**, 17–31. Gallagher's stance was substantially anticipated at Karl Popper (1972), *Objective Knowledge: An Evolutionary Approach*, pp. 69–70.

10 Gallagher, op. cit., pp. 29–30.

is a gradual process, without jumps or 'saltations'.[11] Some of his supporters, however, including Thomas Henry Huxley, have regarded saltations as better explaining e.g. the evolution of the eye;[12] indeed saltationism remains a minority position among some Darwinians (including distinguished ones such as John Maynard Smith).[13] But most regard gradualism as essential to the theory of natural selection, holding that on no other basis can the evolution of biological complexity and adaptedness be explained.[14] To this claim we shall be returning (see Chapter Seven, Section 4 below). Gradualism is consistent with the theory of Gould and Niles Eldredge (whatever these authors may have held to the contrary) that evolution consists of long periods of imperceptible change punctuated by episodes where change is rapid (punctuated equilibria), since rapid change could still be due to a rapid sequence of small variations.[15]

Darwinians also diverge over whether natural selection is the sole cause (or the sole important cause) of most biological phenomena (adaptationism),[16] or whether there are other factors at work. Darwin himself firmly expressed the latter view, in the last edition of *Origin of Species*.[17] The view that there are other important factors has been defended by Gould and Richard C. Lewontin. They argue that just as spandrels (triangular tapering concave surfaces formed by the intersection of two rounded arches at right angles) such as those of St. Mark's Cathedral at Venice bear designs that make them seem central to the architect's purpose, but are in fact inevitable by-products of any building plan involving a dome and arches, so biological phenomena can be by-products of factors other than selection, such as inherited 'building plans' that were adaptive for earlier stages in life or for earlier generations or for ancestral species (such as homologies in bone structures between birds and mammals).[18] Further possibilities are listed; most striking is that of the secondary utilization of parts present for other reasons. (Thus even if blushing is an adaptation for sexual selection in humans, it cannot figure in explanations of why blood is red.[19]) In a later paper, Gould and Elisabeth

---

[11] Ruse, op. cit., p. 23.

[12] Ruse, ibid., p. 25.

[13] John Maynard Smith (1981), 'Did Darwin Get it Right?', *London Review of Books*, **3**.11, 10–11; see also Maynard Smith (1982), *Evolution and the Theory of Games*; and Ruse, op. cit., p. 30.

[14] Ruse, ibid., p. 136; thus Dawkins, *The Blind Watchmaker*.

[15] Niles Eldredge and Stephen Jay Gould (1972), 'Punctuated Equilibria: An Alternative to Phyletic Gradualism', in T.J.M. Schopf, (ed.), *Models in Paleobiology*, 82–115.

[16] Elliott Sober (1998), 'Six Sayings about Adaptationism', in David L. Hull and Michael Ruse, *The Philosophy of Biology*, 72–86, p. 72.

[17] Darwin (1872), *The Origin of Species* (Darwin's last edition), p. 395.

[18] Stephen Jay Gould and Richard C. Lewontin (1994), 'The Spandrels of San Marco and the Panglossian Paradigm: A Critique of the Adaptationist Programme', in Elliott Sober (ed.), *Conceptual Issues in Evolutionary Biology* (2nd edn), 73–90.

[19] Gould and Lewontin, ibid., p. 85.

S. Vrba coin the term 'exaptations' for features that are fit for their current role but not 'designed' and/or selected for it, such as lactation in mammals (granted that lactose-like chemicals probably evolved not for nourishing the young but to kill bacteria) or the supportive role of bones, originally selected as stores for phosphates rather than for rigidity.[20] Nevertheless, as Ruse affirms, Gould and his fellow-writers remain Darwinists, recognizing both evolution from a common ancestry, and natural selection as the most important source of biological adaptation and evolution. Gould has himself recently expounded and defended this generic Darwinism,[21] which can be combined with a rejection of the kind of sociobiological explanations of human behaviour adopted by Dawkins.[22]

Yet other Darwinians combine recognition of natural selection with the view that there is much in the realm of biology that lies beyond its scope, such as change at the molecular level, as studied by Motoo Kimura.[23] Adaptationists reply to their various critics that the very case of anti-adaptationists turns crucially on secondary adaptation and its preconditions, and can thus be reconciled with the centrality of adaptation.[24] Ruse calls this position 'ultra-Darwinism',[25] because it is more Darwinian than Darwin himself was. (We return to the bearing of adaptationism and rival theories on the problem of evil in Chapter Seven, and on the emergence of human culture in Chapter Nine.)

While this debate cannot be resolved here, it has importance for the study of human behaviour and society, which could in theory be closely determined by selection (possibly at the level of genes, as argued by sociobiologists[26]), or may be moulded in large measure by factors such as culture, independently of selection and genetic inheritance. Certainly the theory of Richard Dawkins (and more recently of Daniel Dennett) that cultural patterns not determined by genes are determined by postulated self-replicating entities that they call 'memes'[27] (items such as tunes, theories and religious beliefs, regarded as commandeering and colonizing human brains) has not been found to cohere with the phenomena of

---

20 Stephen Jay Gould and Elisabeth S. Vrba (1998), 'Exaptation: A Missing Term in the Science of Form', in Hull and Ruse, op. cit., 52–71.

21 Stephen Jay Gould (1997), *Life's Grandeur*, p. 138.

22 Richard Dawkins (1976), *The Selfish Gene*.

23 Ruse, op. cit., p. 31; Stephen Jay Gould (2000), 'More Things in Heaven and Earth', in Hilary Rose and Steven Rose (eds.), *Alas, Poor Darwin*, 85–105, p. 92.

24 Sober, 'Six Sayings about Adaptationism'.

25 Ruse, op. cit., p. 29.

26 Thus Edward O. Wilson (1975), *Sociobiology*; Dawkins, op. cit. For criticisms, see Philip Kitcher (1985), *Vaulting Ambition*, and John Dupré (2003), *Darwin's Legacy*, pp. 77–98. Dupré criticizes evolutionary psychology, sociobiology's successor, in Dupré (2001), *Human Nature and the Limits of Science*.

27 Dawkins, *The Selfish Gene*, pp. 206–15, a theory revised in Dawkins (1981), *The Extended Phenotype*, p. 109; see also Daniel Dennett (1995), *Darwin's Dangerous Idea*.

rationality and conscious rule-following in human life.[28] The suggestion that the real agents of history are neither God nor human beings but mindless factors such as genes and, within human culture, memes seems an attempt at reductionism that implausibly smuggles purpose back in at the level of microscopic self-replicators (or, in the case of memes, abstract ones).[29] If behaviour is ultimately determined by genes and memes, then, despite the arguments of these theorists, insufficient scope is left for rationality, autonomy or the freedom to do otherwise. A theory of evolution is needed that facilitates rationality and freedom, rather than conjuring them away. (But it is to Dawkins's credit that he rejects anthropocentrism, both in ethics and in metaphysics, a rejection expressive itself of precisely such rationality and freedom.[30])

At the opposite end of the spectrum of Darwinism, Lewontin has argued that the units of evolution range from cells and individuals through kin-groups to species, and also that evolution by natural selection need not be premised on Malthusian competition; for survival and reproductive success can depend more on opportunistic adaptation and on rates of reproduction than on competition for scarce resources. An implication is that even if Darwinism applies to the realm of culture, Darwinian accounts of human society need not be conceived on the model of evolution from pre-human ancestors engaged in and moulded by a struggle for survival, whether between individuals or between kin-groups; indeed it is equally apt to envisage humanity as constructing much of the environment in which selection currently takes place.[31] (We return to the bearing of Darwinism on human culture and society in Chapter Nine, Sections 2 and 3, and to the relations of humanity to the environment in the final section of that chapter.)

Attempts such as those of Dawkins and Dennett to extend the scope of Darwinian explanations to human culture admit of a further classification of Darwinism, this time in terms of philosophical or metaphysical allegiances. Thus Janet Radcliffe Richards divides understandings of Darwinian explanations into three kinds. One variety combines recognition of evolution by natural selection with what she calls the 'Mind First' view; for this view, God, or a designing intelligence, is still needed as creator and sustainer.[32] Using the disparaging phrase of Daniel Dennett, she makes the role of God (even as creator) sound interventionist, speaking of such explanations as 'skyhooks', as opposed to

[28] Mary Midgley (1979), 'Gene-Juggling', *Philosophy*, **54**, 438–58, pp. 456–8; Midgley (2000), 'Why Memes?', in Hilary Rose and Steven Rose (eds), *Alas, Poor Darwin*, 67–84; David Holdcroft and Harry Lewis (2001), 'Consciousness, Design and Social Practice', *Journal of Consciousness Studies*, **8**.8, 43–58.

[29] Keith Ward (1996), *God, Chance and Necessity*, pp. 138–9, 166–9.

[30] Dawkins, *The Blind Watchmaker*, pp. 262–3.

[31] Richard Lewontin (1970), 'The Units of Evolution', *Annual Review of Ecology and Systematics*, **1**, 1–23; see also Daniel Philip Todes (1989), *Darwin Without Malthus*.

[32] Janet Radcliffe Richards (2000), *Human Nature after Darwin*, pp. 54–5.

the earth-bound, material explanations or 'cranes' exemplified by Darwinian explanations.[33]

The other two varieties are both 'Matter First' views which reject the 'Mind First' view on a materialist basis, and invoke Darwinian-like, non-teleological explanations not only for apparent design in biology and for the origins of life but also (in some cases) for human culture and society; some even seek such explanations for the inanimate universe too. They differ where one kind holds that evolutionary origins can tell us little about minds, meanings and society, something that from this perspective requires the methods of social science (what she calls the 'blank-paper view'), whereas the other variety believes that psychological and social phenomena are to be explained by an extension of Darwinian explanations via evolutionary psychology, with genetic factors and constraints taking centre-stage (what she describes as the 'gene-machine view').[34] (Thus the terms used by Richards for all these stances are ones their critics might give rather than their advocates.) Between these stances there are ongoing skirmishes, which Andrew Brown has called the 'Darwin wars'.[35] Richards's classification would be enhanced (and better fit the range of occupied positions) by recognizing, among 'Matter First' theorists, agnostics such as Gould, as well as atheists (like Dupré). But the label 'blank-paper view' so considerably misrepresents the stance of theorists such as Gould and Lewontin that I will instead refer of them on appropriate occasions as holding the 'moderate materialist view'.

Richards's classification of Darwinisms (or what might be called 'meta-Darwinisms') will prove useful in what follows, alongside the distinctions between adaptationism and the various schools of its Darwinian critics. Further, different views about the scope of Darwinism (mentioned above) become important when its bearing on meaning, language and culture are considered in Chapter Nine. Creationists and their allies such as Plantinga, meanwhile, reject all these stances; while some of them accept evolution of a non-Darwinian kind, they usually reject common ancestry and natural selection. Arguments of Plantinga and others for the incompatibility of Darwinism and Christianity are discussed in the course of the two coming sections.

---

33 Daniel Dennett, op. cit., pp. 73–7. The foundational passage that introduces cranes and skyhooks (ibid., p. 75) comes as close to a definition of his metaphor of cranes and skyhooks as Dennett seems to supply: 'Cranes can do the lifting work our imaginary skyhooks might do, and they do it in an honest, non-question-begging fashion. Skyhooks are miraculous lifters, unsupported and insupportable. Cranes are no less excellent as lifters, and they have the decided advantage of being real.'

34 Richards, op. cit., pp. 55–6.

35 Andrew Brown (1999), *The Darwin Wars*.

Section 2: Darwinism, Theism and Gould

It is time to consider the mutual bearing of Darwinism and theism. Are they compatible, or alternatively do the grounds for either of them undermine the grounds for the other? In particular, does Darwinism undermine the Design Argument for the existence of God in any or all of its varieties? It certainly seems to undermine the variety of Design Argument put forward by William Paley, since it better explains the adaptations that Paley cites, and thus makes his watch-maker explanation redundant. But as Alister McGrath remarks, Paley's entire approach had already been disowned before the publication of Darwin's great book by such leading theologians as John Henry Newman and R.S.S. Baden-Powell, the latter commending a search for biological laws to explain (what we call) adaptations, rather than invoking a designer for each of them one by one.[36]

The suggestion that Darwinism discredits theism is implicit in the work of Dawkins and of Dupré, since (among other considerations) it supposedly makes the only cogent argument for theism (Paley's) obsolete.[37] There are also problems for reconciling the randomness that is at the heart of the Darwinian synthesis with divine design, and the ruthlessness that it seems to ascribe to nature with the goodness of God. However, strong grounds for both theism (in Chapter Five) and Darwinism (in the current chapter) have been exhibited above, and thus the suggestion that they cannot both be true (or that either could in the circumstances readily be considered untrue) confronts serious problems. In the case of theism, the Cosmological Argument was found to support key aspects of theism, while at least one variety of the Design Argument was argued to support some of its other crucial aspects, and not to have been undermined by the criticisms of (at least) Hume.

As for Darwinism, the grounds presented by Darwin prove impressive, and have been supplemented in the century and a half since *The Origin of Species* was published by the phenomena of Mendelian genetics and further evidence from geographical distributions of species. The various grounds (from field and laboratory experience, as well as from homologies and from the fossil record) have been well marshalled by Stephen Jay Gould, and become all the more formidable if their conclusion is not unqualified adaptationism (see the previous section) but the more historically Darwinian claim that natural selection is one important factor (among others) in the explanation of biological phenomena, indeed the most important.[38] The Christian biologist R.J. Berry well defends the evidence of the fossil record against creationist objections.[39] Not even the scepticism of such an able critic as Plantinga (and his view that Gould's 'punctuated equilibriumism'

[36] Alister McGrath (2005), *Dawkins' God*.

[37] Dupré, *Darwin's Legacy*; Dawkins, *The Blind Watchmaker*.

[38] Stephen Jay Gould (1983), 'Evolution as Fact and Theory', in Gould, *Hen's Teeth and Horse's Toes*, 253–62, pp. 257–9.

[39] R.J. Berry (1988), *God and Evolution*, pp. 106–7.

amounts to the Darwinian equivalent of a Ptolemaic epicycle[40]) suffice to weaken the strength of Gould's variety of Darwinian stance. The possibility certainly remains, for all that we have seen so far, that Darwinism may somehow supplement Hume's criticisms and undermine the Design Argument altogether, but even if this were plausible, it could hardly undermine the Cosmological Argument as well. By the same token, Darwinism can hardly establish materialism, and strictly 'Matter First' versions of Darwinism (like those of Dawkins and Dennett) are likely to prove unstable, at least by comparison with agnostic versions that refrain from asserting that theism is either false or insupportable.

The case should therefore be considered for the compatibility of Darwinism and of theism, and, if this case can be sustained, the related case for the cotenability of Darwinism and the kind of belief in God and in divine design that Darwin endorsed during the middle period of his life. Two recent Darwinian writers have produced book-length arguments seeking to reconcile the stances of Darwinism and theistic religion, an agnostic, Stephen Jay Gould, and an ex-Quaker, Michael Ruse.[41] While neither work is above criticism, each produces powerful reconciling arguments.

The core of Gould's case is that the realms of authority (magisteria) of science and religion are distinct and non-overlapping; hence his slogan 'NOMA' (non-overlapping magisteria).[42] The realm of science is variously characterized as that of explanations of observable phenomena and that of factuality, while the realm of religion is regarded as that of ethics and of the meaning of life; and there is this much to be said in favour of his stance, that science is competent to deal neither with ethical issues nor with issues of what makes life meaningful, while religion is incompetent to deal with how nature operates and with the field of observable recurrent natural phenomena. Gould argues that both Darwin and his supporter Thomas Henry Huxley recognized this same stance,[43] but it is unnecessary to follow the details here. One problem for this stance is the possibility that there are ethical truths, and thus ethical facts; but Gould's treatment of the fact/value distinction, about which he claims no expertise,[44] can also be set on one side. So can the issue of whether religious ethics has any more central role in the realm of values than secular ethics; Gould's point here is that religions have historically embodied ethical teaching, and continue to do so, even though there is a separate issue of whether or not to recognize religious authority in this area.[45]

---

40 Plantinga, 'When Faith and Reason Clash', pp. 683–92.

41 Stephen Jay Gould (1999), *Rocks of Ages*; Ruse, *Can a Darwinian Be a Christian?* For an argument from a theistic perspective for the compatibility of Darwinism and theistic interpretations of the universe, see John Cottingham (2003), *On the Meaning of Life*, pp. 44–62.

42 Gould, ibid., p. 52.

43 Gould, ibid., pp. 27–45.

44 Gould, ibid., pp. 54–7.

45 Gould, ibid., pp. 56–60.

Far more problematic for Gould's stance is his apparent exclusion of religion from the realm of factuality. Dupré draws to attention the problem that if all ontological claims are denied to religion, then the compatibility of science and religion is purchased at a price too heavy for the claims of most religions. The waters are muddied here by Dupré's implausible claim that science has through Darwinism undermined any plausible grounds for theistic beliefs.[46] However, Dupré's central criticism is on target, and Gould's explication of NOMA needs to be revised such that science is concerned not with factuality in general (a view that would emasculate history and other humanities disciplines as well as religion) but with (say) the field of observable recurrent natural phenomena and natural explanations of them. Natural science (including evolutionary theory) is not concerned with non-recurrent and naturally inexplicable phenomena such as miracles (if there are any), nor centrally with past human actions and their interpretation, nor standardly (unlike social science and the humanities) with human culture and its interpretation,[47] nor with meta-scientific issues such as the existence or non-existence of a creator.

Thus reinterpreted, and equipped with qualifications recognizing interfaces between scientific and religious concerns, Gould's stance could, I suggest, be made defensible. Theistic explanations would not be excluded altogether, as in the kind of metaphysical naturalism contested by Plantinga,[48] but would, *pace* Plantinga, be debarred *within science* (as in the 'methodological naturalism' commended by Ruse[49]), to avoid the kind of 'physical theology' or invocation of supernatural explanations for particular physical phenomena wisely rejected by theologians such as Newman and Austin Farrer.[50] Agnosticism (such as that of Gould) would certainly be no obstacle to endorsing such a stance. But Gould's implicit condemnation of theistic interpretations of the laws of nature, in the forms of the Fine Tuning Argument and also of the version of the argument to a designer from regularities of succession,[51] as being precluded by NOMA,[52] is an unnecessary obstacle to his project of reconciliation, and would need to be retracted. Hence a satisfactory reconciliation would need to adopt a rather different form from that envisaged by Gould. But at this juncture, Ruse's parallel project also needs to be considered.

---

46 Dupré, *Darwin's Legacy*, pp. 57–8.

47 R. Hooykaas supplies grounds for this view in Hooykaas (1974) 'Nature and History', in Open University Course Team for Science and Belief: from Copernicus to Darwin, *The New Outlook for Science*, 10–33, pp. 20–23.

48 Plantinga, 'When Faith and Reason Clash', pp. 693–5.

49 Ruse, *Can a Darwinian Be a Christian?*, pp. 99–110. See also Ruse (1995), *Evolutionary Naturalism*, pp. 2–4.

50 For Newman, see McGrath, op. cit., pp. 67–9; for Farrer, see A.M. Farrer (1952), 'Introduction' in G.W. Leibniz, *Theodicy*.

51 For the Fine Tuning Argument, see Chapter Five, Section 5 above; for the argument from regularities of succession, see Chapter Five, Section 4 above.

52 Gould, *Rocks of Ages*, pp. 219–20.

Section 3: Darwinism, Theism and Ruse

For his part, Michael Ruse investigates several areas where Darwinian and Christian beliefs are held to conflict (such as those of purposive explanations and of the origin of life), and concludes in each case that there is no contradiction. He also replies to the arguments of Plantinga that the methodological naturalism (or insistence on naturalistic explanations as a requirement of scientific method) implicit in Darwinism (and also in the science of the scientific community in general) inclines its adherents towards metaphysical naturalism and thus to atheism and materialism. According to Plantinga, science either rejects explanations that are independent of scientific laws or relegates them to the status of at best second-class knowledge, and thus cannot cope with non-repeatable occurrences (which would in practice be unfalsifiable); yet God is free to act independently of natural laws, and Christianity turns on non-repeatable events. An unbiased form of science would have to discard methodological naturalism and the falsifiability requirement that characteristically goes with it, and recognize divine interventions as well as divine laws; Plantinga calls such a science 'Augustinian science'.[53] However, as Ruse replies, science need reject neither belief in creation nor in miracles, as opposed to regarding both as beyond its scope. It can retain the methodological naturalism that has served it so well, together with the associated falsifiability criterion, and the related rejection, for purposes of the field of scientific investigation, of supernatural explanations, miracles and non-repeatable events, without its adherents having to hold that there are no supernatural explanations, non-repeatable events or miracles. For scientists need not hold that science is the sole form of investigation or of reason.[54] Even if some Darwinians subscribe to the materialist range of Richards's spectrum, there is nothing to prevent Darwinians being either agnostics (like Gould) or religious believers (like Fisher and Dobzhansky).

Plantinga has also suggested that Darwinism implies that rational beliefs are all to be explained by naturalistic factors and forces such as the need to survive and reproduce, and that for such an approach even our most rational beliefs need not be true or reliable, and could be deceptive or delusive. Thus naturalism is self-refuting.[55] To this Ruse replies that evolution has given us reliable touchstones of deception and delusion, and thus the ability to discriminate between cases of deception and cases of non-deception. Thus Darwinists as well as their opponents can take truth seriously, and aspire to attain it.[56] To this reply it should, I suggest, be added that Plantinga's objection may well be on target against consistently

53 Plantinga (1997), 'Methodological Naturalism', *Perspectives on Science and Christian Faith*, **49**.3, 143–54.

54 Ruse, *Can a Darwinian Be a Christian?*, pp. 99–106.

55 Alvin Plantinga (1993), *Warrant and Proper Function*, pp. 218–33.

56 Ruse, op. cit., pp. 106–9. See also the replies to Plantinga of Willem B. Drees (1996), *Religion, Science and Naturalism*, pp. 156–8.

deterministic varieties of ultra-Darwinism, for which our beliefs exist simply as a result of natural selection and other factors beyond our control, and are the beliefs that we could not help adopting in the circumstances; if so, the project of attaining the truth would often be illusory as well as dysfunctional, and would have slender prospects for survival. But Darwinism can instead be held in an indeterminist or libertarian version that allows room for the exercise of self-determination and (ultimately) rationality;[57] and it is this kind of Darwinism (and perhaps only this kind), one that provides for rational discrimination and discretion, that is defensible along the lines of Ruse's reply, and that can thus fend off the criticisms of Plantinga. (This defence is not strictly available to Ruse, who subscribes to the compatibility of determinism and human freedom, and is a determinist in the matter of 'hard-wired' desires.[58]) Given that Plantinga's objection seems well taken against deterministic varieties of naturalism, it is of great importance that neither Darwinism nor naturalism in general need be deterministic.[59]

Ruse in fact holds that Darwinism and Christianity are in many ways mutually supportive.[60] Such a stance had earlier been adopted by Keith Ward, who concludes that 'There is thus every reason to think that a scientific evolutionary account and a religious belief in a guiding creative force are not just compatible but mutually reinforcing'.[61] Ruse's discussions of issues relating to suffering and concerning alternative natural systems to that of the actual universe raise problems and will be revisited when we reach issues of value and disvalue. Dupré finds Ruse's argument for the compatibility of Darwinism and Christianity generally successful (and with good reason), but again objects (this time 'in disconcerting accord with Richard Dawkins and Christian fundamentalists') that Darwinism clashes with Christianity by undermining 'the only remotely plausible reason for believing in the existence of God'.[62] Besides being unfair to creationists (who often have other grounds for belief), and completely ignoring the Cosmological Argument, Dupré here assumes that if Paley's argument fails, so does the Design Argument in all its forms.

But this assumption is mistaken. Darwinism certainly supplies a natural explanation for the adaptations to which Paley appealed, and for some of the phenomena of apparent design discussed by Hume; sceptics about Paley's

---

57 For the notion of indeterministic or libertarian freedom, see further Chapter Eight, Section 1 below; also Keith Ward (2004), 'Theistic Evolution', in William Dembski and Michael Ruse (eds), *Debating Design*, 261–74, p. 263.

58 Ruse, op. cit., pp. 205–16.

59 Relatedly, the moderate Darwinist Steven Rose explains how living systems should not be regarded as biologically determined, but continually construct their own futures; see Rose (1997), *Lifelines: Biology, Freedom, Determinism*, pp. 5–7 and 18. See further Chapter Eight, Section 5 and Chapter Nine, Section 4 below.

60 Ruse, *Can a Darwinian Be a Christian?*, e.g., p. 137.

61 Ward, *God, Chance and Necessity*, pp. 63.

62 Dupré, op. cit., p. 56.

argument no longer needed to appeal to theoretically possible explanations (such as Hume's thought that given long enough anything could turn up, including apparent contrivance), once an explanation based on facts about fossils, geography, population dynamics and the phenomena of selection became available. Darwinism may also be capable of explaining cases of supposedly irreducible complexity (such as the visual system comprising the eyes and brains of higher animals),[63] put forward as counter-examples by creationists and advocates of Intelligent Design. But the argument from the regularities of the natural order across time to a designer of these regularities remains unaffected, as does that from Fine Tuning. As was mentioned in Chapter Five, while Darwinism makes the adaptedness of animals to survival unsurprising and thus helps to explain the survival of the relevant adaptations, it presupposes that the world's processes are regular enough for such adaptations to have emerged and to have persisted, and takes for granted a context of universal laws (such as those of gravity, light, sound, heat, and electro-magnetism) but for which the world could easily have been vastly different and/or much less regular than it is. Indeed, laws governing natural selection amount to further instances of such laws.

But these regularities across space and time stand in need of explanation, much as diachronic regularities of individual or social human contrivance do (such as symphonies, dances or songs, or, differently, schedules and timetables of recurrent events). Such natural regularities cannot evolve, or thus be explained along Humean or Darwinian lines. Whether or not the Principle of Sufficient Reason is accepted, these striking, all-pervasive regularities can only be explained by the activity of something with commensurate power and intelligence. While it is possible to consider this argument an argument from analogy, it could instead be regarded as one from rational interpretation, depending not so much on likenesses between symphonies, schedules and the universe, as on its conclusion being the best explanation of the regularities in question. Such reasoning is not physical theology, which brings in supernatural explanations for particular phenomena alongside natural explanations; rather it seeks to explain contingent but universal properties of the actual universe which could have been otherwise by reference to an agent capable of conferring these or other sets of universal properties on the created order. Much the same goes for the argument from Fine Tuning, understood as an argument to the best explanation of universal constants embedded in laws but for which life would be impossible. This version of the Design Argument too stands unaffected by Darwinism.

These arguments (as thus presented) say little or nothing of the character of the designer (except that the designer of the regularities argument favours order and that of the Fine-Tuning Argument also favours life); as thus presented, they are largely or entirely independent of such considerations (and thus of objections about waste and suffering too). Nor do they require the designer to be omnipotent, omniscient, uncreatable or indestructible. But their conclusion is consistent

---

63 See Dawkins (1996), *Climbing Mount Improbable*, pp. 126–79.

with the designer's being identical with a person or being having such attributes (see Chapter Five, Section 4 above), and this is particularly significant if other considerations (such as the Cosmological Argument) support the existence of such a being (as was argued in Chapter Five, Section 2). If this identity holds, then no questions arise about how to explain the existence of the designer of cosmic regularities and/or of ones fine-tuned for the existence of life. For questions about the reason for the existence of a creator with such attributes (with whom the designer is taken to be identical) cannot be asked and do not even make sense. (See Chapter Four, Sections 3 and 4 above.)

The randomness on which natural selection depends may still seem a problem; but a designing intelligence might choose to create life through a process with such a basis; so the woodpecker's tail need not be understood by theists to be specifically designed for its purpose. (The issue of randomness is revisited in the final section of Chapter Seven.) There again, the ruthlessness that Darwinism seems to ascribe to nature would be a problem for some kinds of theistic belief. But belief in natural selection need ascribe ruthlessness neither to individual creatures nor to ecosystems, let alone suggest that being ruthless forms the best route to survival.[64] To such topics (and the related topics of evil and suffering) we shall return, both in this and the next chapter. However, the interim conclusions are that Darwinism and theism are compatible and cotenable, that Darwinism does not undermine all the varieties of the Design Argument, let alone the grounds for theism, and that good grounds remain for endorsing theism as well as Darwinism.

Section 4: Disvalues in Nature

Some of Darwin's grounds for questioning theism remain to be considered, grounds relating to suffering and evil. In the previous century, Hume had more explicitly presented arguments, through the mouth of his character Philo, against God's goodness, suggesting ways in which the universe could have been made better. In this section, disvalues in nature and related claims and comments of Hume and of Darwin are discussed, and in the coming section some more recent interpretations are introduced and sifted. Arguments surrounding the bearing of evil on God's goodness are considered more thoroughly in Chapter Seven; this chapter is confined to Humean and Darwinian considerations.

In 1860, Darwin wrote as follows to his friend the Christian biologist Asa Gray: 'I cannot persuade myself that a beneficent and omnipotent God would have designedly created the Ichneumonidae with the express intention of their feeding within the living body of caterpillars.'[65] These insects, whose eggs are

---

64 Ward., op. cit., p. 64.

65 Darwin, letter to Asa Gray, 22 May 1860, Francis Darwin (ed.) (1888), *The Life and Letters of Charles Darwin*, vol. 2, 310–12; cited in Richard Dawkins (1995), *A River Out of Eden*, p. 95.

laid inside hosts that the young creatures then consume, serve as an example of parasitism, which is widely considered a defect of the natural order. The issue is more than whether this practice is directly designed (the issue discussed above about the woodpecker's tail). It concerns suffering and waste that cannot readily be reconciled with the purposes of a benevolent creator.

For Darwin, human suffering was also an obstacle to Christian belief. In addition to his own protracted illness, Darwin was particularly affected by the loss of his daughter Annie in 1851 at the age of ten. He seems to have eventually concluded that 'pain and suffering were to be accepted as the meaningless outcome of the evolutionary process', rather than that God permits or authorizes their infliction.[66] Already by 1860 he held that suffering is a chance product of designed laws;[67] by 'chance', as Gould explains, he probably meant not 'random' but 'contingent and unplanned'.[68]

Hume approached these matters more systematically. Following Epicurus and Voltaire, his character Philo asks why, if God is good, evil is still to be found. God's omnipotence would mean that he is not unable to prevent it, and his omniscience that he does not lack the necessary knowledge or skill. While some evils may be necessary, four possible ways are advanced by Philo in which there could have been a better world. The fact that they are not in place in the actual universe suggests that there is not a benevolent designer, or at least that there cannot be known to be, for the existence of avoidable evil is (supposedly) incompatible with there being an all-powerful, all-knowing and benevolent creator.[69] (This incompatibility argument will be addressed in Chapter Seven, Section 1.)

Hume interprets evils as 'ills that molest sensible creatures',[70] and this can serve as a temporary definition, although it leaves unresolved issues such as whether evils can befall insentient creatures such as trees, whether failure to flourish is sometimes an evil, and whether evils include extinctions of species, irrespective of individual harm or loss. His first suggestion for a better world is that pain is unnecessary, and that its functions could be served by diminution of pleasure.[71] This is at least a proposal for a better world-system, and therefore, as will be argued below, a suggestion of the appropriate general kind. But this suggested alternative to pain is hardly satisfactory, since the warning function of pain depends on the affected creature's unavoidable awareness of pain, a phenomenon characteristically lacking from reduced levels of enjoyment. Although some pain is dysfunctional, Philo fails here to present a way in which universally recognized functions of pain could be better served, and thus in which the world could have

---

66 McGrath, *Dawkins' God*, p. 74.

67 Gould, *Rocks of Ages*, pp. 35–6.

68 Gould, ibid., pp. 198–9.

69 Hume, David [1779], 'Dialogues concerning Natural Religion', Dialogues X and XI, in Richard Wollheim (ed.) (1963), *Hume on Religion*, 99–204, pp. 165–88.

70 Hume, Dialogue XI; Wollheim, p. 179.

71 Hume, ibid.,Wollheim, pp. 179–80.

better served the purposes of a benign designer. What those purposes might be held to consist in will be further discussed in the course of the next two chapters; but in the current context, Hume's criterion can reasonably be taken to be provision for the well-being or flourishing of sentient creatures.

Hume's second suggestion, that the world would be better if governed not by general laws but by 'particular volitions' of the author of nature,[72] is of a different kind, and his other two suggestions should be considered first. His third suggestion is that the distribution of powers of animals should be less 'frugal', since at present they barely have sufficient powers for survival. Darwinians can scarcely sympathize with this account, holding as they do that survival is due to differential capacities of fitness to survive and reproduce in the relevant environment. Hume comments that human beings survive only through their skill and industry; but these are the very powers that have enabled humanity to colonize most of the terrestrial environments of the planet.

Nevertheless Hume proceeds to illustrate his general point by suggesting that the world would be better if humanity were less indolent, and endowed with 'a greater propensity to industry and labour' and with the application to carry out what they resolve. Hume could counter the *ad hominem* reply that his readers must have such application if they had reached Dialogue XI, since he restates his suggestion as favouring humans being endowed with 'an equal diligence with that which many individuals are able to attain by habit and reflection'.[73] But this remark serves to remind us that application and industriousness are habits acquired through effort and earlier choices, rather than through heredity. Hence if some human beings are deficient in these regards, this is a defect within human control, and not to be ascribed to God or nature or heredity. People develop the qualities favoured by Hume through habits formed early in life, the encouragement of others, and responses to such encouragement, or in other words through free human activity; this being so, Hume's example fails to epitomize his claim that the inherited powers of creatures should have been ampler. (The different objection that a creator would have ensured that the outcomes of free choices were better than they are is considered in the coming chapter.)

Hume's fourth suggestion is that the components of our natural system (winds, rain, heat and human passions are his examples) often fail to perform their function, whether by excess or deficiency; if they are designed, they are inaccurately crafted.[74] If human convenience were the criterion of functionality, this suggestion would be understandable, granted the destructive power of hurricanes and tsunamis. But when the criterion is taken to be belonging to a system of natural laws supportive of life, then winds, rain and heat conspire within a system that functions well. (If things were otherwise, the Fine Tuning Argument, as discussed in Chapter Five, could never get started.) What is to be feared is

---

72 Hume, ibid., Wollheim, pp. 179–80.

73 Hume, ibid., Wollheim, p. 183.

74 Hume, ibid., Wollheim, pp. 184–5.

anthropogenic (humanly originated) disruption of the component ecosystems of the system of nature (for example, anthropogenic global warming), and not the unmodified operation of the system itself. (Hume's earlier suggestion that laws of nature are part of the problem will be considered shortly.) As for human passions (and the humours and juices of the body that Hume takes to underlie them),[75] Darwinians can hardly deny that they serve the overall functions of survival and reproduction. And although, as Hume remarks, they can also lead to individual ruin, such cases are avoidable through the free choices either of the person concerned or of others, if not of both, and so the complaint should not be directed at components of the system of nature.

To return now to Hume's second suggestion, it is Hume's view that the operation of any system of laws of nature is bound to lead to casualties.[76] Two scenarios are considered in an attempt to depict a better world. In one, there are no laws of nature, 'the course of nature' is 'perpetually broken', and 'no man could employ his reason in the conduct of life'. Hume recognizes this 'inconvenience', and thus proposes a second scenario as preferable. But (where the first scenario is concerned) the absence of natural regularities and the inapplicability of reason are far more than an 'inconvenience'; not only would there be no science and no scope for rationality, there would be no creatures of the kinds that have evolved by natural selection within the framework of the laws of nature to which we are accustomed, and if there were any life at all, it would be unrecognizably different, with no recognizable goods or ills remaining. For conscious life, indeed, some system of laws of nature not too different from that of the actual world seems essential.

Hume's second scenario avoids such disorder by reinstating laws of nature, except that the deity exterminates all evils and produces all goods 'without any preparation of long progress of cause and effects'. God is to do this by stealth, 'without discovering himself in any operation'; human ignorance of causes is such that felicitous accidents are never diagnosed as the miracles that they are. But the claim that God's interventions would go unnoticed is entirely implausible. Besides, in such a dispensation, the ordinary course of nature could not be understood at all, as effects that were otherwise predictable would fail to materialize whenever good or evil was at stake. Even if these interventions were restricted to cases where gross evils were removed (a possibility implicit in Hume's suggestion that 'some small touches to Caligula's brain in his infancy might have converted him into a Trajan'),[77] the ordinary course of nature could still not be understood, since impending calamities would always mysteriously fail to happen (and even efforts to make them happen, including acts of war and of terrorism, would mysteriously miscarry). There would thus be no regular natural sequences to be observed, and so no possibilities for human science or technology would remain. Besides, if the

---

75 Hume, ibid., Wollheim, p. 184.

76 Hume, ibid., Wollheim, p. 182.

77 Hume, ibid., Wollheim, p. 181.

good of all creatures were guaranteed, there would be no natural selection, and thus no evolution of species, unless that too were generated by different laws, or miraculously introduced by acts of special creation. If, however, intelligence, freedom, rationality and creativity are among the goods that the author of nature wanted creatures to attain, then a world order unlike either of Hume's scenarios (and much more dangerous than his second one) would be created, one suited to the emergence of these qualities, and made understandable through the operation of natural laws, subject to scope for the free choices of creatures.

Hume maintained that most natural evil is due to the universe not complying with his four suggestions,[78] where natural evils are evils not subject to human agency or control. But in fact, as has been seen, he failed to depict ways in which a better universe could have been brought about. The suggestion made here is emphatically not that there is no natural evil, but rather that the evils associated with Hume's suggestions mean that they would not be improvements. His suggestion of a world operating through frequent (or even perpetual) miracles would produce a disordered universe incomprehensibly different from the setting in which intelligent species have, and could have, evolved;[79] while his suggestions for a world with different and better regular properties do not survive scrutiny. Some further discussion of suggestions of this type can be found in Chapter Seven.

Unlike Hume, Darwin accepted universal laws of nature; indeed his theory of evolution supplemented our understanding of such laws. But whether or not such an order of nature, Ichneumonidae and all, is consistent with benevolent design (something Darwin came to doubt) remains to be further considered.

Section 5: Nature and Immorality

Darwin's passage on the Ichneumonidae has received widespread discussion among Darwinists of different persuasions; in this section, some interpretations of disvalues in nature from followers of Darwin are introduced and sifted. Gould, for example, depicts some of the glosses of cosmic optimists on moral lessons to be learned either from the mother wasp's considerateness for her young or the self-discipline of the grubs,[80] to which he appends a fitting parody from Mark Twain.[81] Gould's own view, that nature is amoral and enshrines no such lessons,[82] appears acceptable, until he turns out to be saying that nothing in nature has intrinsic value either.[83]

---

78 Hume, ibid., Wollheim, p. 185.

79 See further Bruce Reichenbach (1982), *Evil and a Good God*, pp. 87–110.

80 Gould, *Rocks of Ages*, pp. 183–7.

81 Gould, ibid., pp. 187–8.

82 Gould, ibid., p. 195.

83 Gould, ibid., p. 178.

Some of Darwin's followers (not only from the Social Darwinist tendency) have, however, concluded that natural creatures, and even nature itself, should be declared immoral, and some have deplored the restraint of Darwin's early supporter Thomas Henry Huxley, when he wrote: 'Thus, brought before the tribunal of ethics, the cosmos might well seem to stand condemned. The conscience of man revolted against the moral indifference of nature, and the microscopic atom should have found the illimitable macrocosm guilty.'[84] Huxley has recently been reproached by the biologist George C. Williams for excessive moderation, and for exhibiting the English tendency to understatement. This is what Williams, writing in 1988, would prefer him to have said: 'Thus, brought before the tribunal of ethics, the cosmos stands condemned. The conscience of man must revolt against the gross immorality of nature.'[85]

Williams certainly endorses Huxley's view of social progress, presented by Huxley as follows: 'Let us understand, once and for all, that the ethical progress of humanity depends, not on imitating the cosmic process, still less on running away from it, but in combating it',[86] and reasonably enough, provided that at the same time we remember, with the anarchist biologist Petr Kropotkin and the sociobiologist Frans de Waal, that there is actually much intra-specific co-operation (so greatly needed in human affairs) in nature too.[87] But Huxley's belief in the moral neutrality of nature is claimed to be an error of judgement by Williams, who goes on to claim that 'The inescapable arithmetic of predation and parasitism should be enough to show that nature is morally unacceptable',[88] and holds in general that 'what is, in the biological world, normally ought not'.[89]

On the basis of research in sociobiology, Williams is prepared to represent non-human animals as practising slavery, adultery, rape and 'other sins', including deception (as opposed to fidelity), cannibalism, infanticide and nepotism, quite apart from predation and parasitism.[90] He is aware of the objection that human practices, such as slavery, are largely determined by culture rather than genes and differ in many details from anything practised (for example) by ants, and so the term 'slavery' should not be used of ant behaviour. But he is unimpressed by

---

84 Thomas H. Huxley (1894), *Evolution and Ethics and Other Essays*, p. 59. 'Evolution and Ethics' was first presented as his Romanes Lecture in 1893. Much of the current section is based on Robin Attfield (2000), 'Evolution, Theodicy and Value', *The Heythrop Journal*, **41**.3, 281–96.

85 George C. Williams (1988), 'Huxley's Evolution and Ethics in Sociobiological Perspective', *Zygon*, **23**.4, 383–407, at p. 384.

86 This passage of Huxley, 'Evolution', is cited at the head of Williams's paper, at p. 383. It also occurs in T.H. Huxley and Julian Huxley (1947), *Evolution and Ethics*, at p. 82.

87 Petr Kropotkin [1902] (1955), *Mutual Aid*; Frans de Waal (1996), *Good Natured: The Origins of Right and Wrong in Humans and Other Animals*.

88 Williams, op. cit., p. 398.

89 Williams, ibid., p. 383.

90 Williams, ibid., pp. 389–97.

such objections, since they are, to his reckoning, never made against the use by biologists of terms normally used of culture-laden activities such as 'courtship', 'migration' and 'fasting', and he rejects the selectivity implicit in the objection to the use by biologists of such terms only when they connote wickedness.[91]

There are, of course, problems concerning the analogical use in non-human contexts even of terms such as 'migration'. For while both humans and non-humans sometimes traverse large distances *en masse*, human migration is usually understood as migration by the participants, who, when they migrate, are complying with a concept and (at least in principle) could have acted otherwise, whereas there is little or no reason to ascribe these characteristics to the migrations of birds or wildebeests, which is why terms such as 'instinct' are often invoked in such cases. So the senses of predicates such as 'migrate' have to be understood as adjusted to the natures of the creatures to which they are applied (a kind of adjustment implicit in the theory of analogy introduced in Chapter One, Section 6). But as long as this is understood, there seems every reason to continue using the language of 'migration'.

However, terms that connote wickedness or culpability when they are used of human behaviour are different. With infanticide, for example, the human beings concerned are not only capable of understanding what they are doing but also (in normal cases) capable of acting otherwise; against this background, ascriptions of infanticide necessarily convey blameworthiness, at least to some degree. Since this is completely inappropriate in the generality of cases concerning animals, the language of infanticide and of immorality is entirely out-of-place there, as Frans de Waal has remarked, adding that Williams's whole position about animal morality 'unravels' in consequence.[92] Such language certainly remains out-of-place unless it is given a new sense in which wickedness and culpability play no part.

Williams, however, has no desire to retract connotations of wickedness, since it is his contention that nature in its entirety is grossly immoral. Indeed Williams sometimes argues from the immorality of animals to that of the processes of nature as a whole.[93] But his conclusions have been contested by Holmes Rolston, who urges reflection on the overall evolutionary system of nature, and finds considerable 'systemic value' there.[94] ('Systemic value', while distinguished by Rolston from intrinsic value, diverges from instrumental value in that the systems that carry it are no more instrumental than the planet itself is for support of terrestrial life, and yet such life is unthinkable without them. We return to systemic value in Chapter Eight, Section 3.) At issue is the interpretation to be put on phenomena such as predation and parasitism.

---

[91] Williams, ibid., p. 389.

[92] de Waal, op. cit., p. 221, n. 15; for a comparable response, see Holmes Rolston III (1999), *Genes, Genesis, and God*, pp. 83, 275.

[93] George C. Williams (1996), *Plan and Purpose in Nature*, p. 159.

[94] Holmes Rolston III (1992), 'Disvalues in Nature', *The Monist* **75**, 250–78; Section 13 (pp. 275–6) is actually entitled 'Systemic value'.

Some thinkers who apparently deplore parasitism, such as Darwin's younger contemporary George Romanes (1848–1894),[95] have really been objecting to predation, and seem to be saying that there are too many heterotrophs, which feed on other creatures, as opposed to autotrophs, which synthesize their own food, as most flora do. 'Could we have a world with only flora, no fauna?' asks Rolston in response.[96] His answer is 'Possibly not, since in a world in which things are assembled something has to disassemble them for recycling.'[97] Perhaps the truth here is that a world of nothing but flora is logically possible, but that if it were created and provided with laws of nature at all similar to those of the actual world, it would be very shortlived in the absence of herbivores and of funguses (which are not strictly flora), and that a sustainable world of such a character but consisting only of flora is impossible.

The real question, however, is one of value. Would a world consisting solely of flora be better or more valuable than one consisting of both flora and fauna? Rolston asks this too, and comments that no one thinks the answer is 'yes'. 'Heterotrophs must be built on autotrophs, and no autotrophs are sentient or cerebral.'[98] Thus if we value the capacities for sentience and intelligence, or any such sophisticated capabilities, then it seems that we have to favour predation. (Thomas Hardy wrote a poem 'Before Life and After', bewailing the birth of sentience in the natural world; but, as Brian Hebblethwaite comments, such a negative stance is hard to sustain, in view of the 'almost infinitely varied world of animal life with all its energy, vitality and beauty'.[99])

Or rather, we have to do so with things as they are, no autotrophs being sentient or cerebral. But could there have been sentient or cerebral autotrophs? Or could there have been worlds of just plants and plant-eating fauna only (whose grazing comprises the least problematic kind of predation), as opposed to the fauna usually picked out as 'higher predators' (that is, flesh-eaters)? Cerebral autotrophs such as ents and triffids have been imagined for worlds of fiction, but seem likely to be dysfunctional in any non-fictional world. But let us imagine that worlds of the sort suggested a moment ago remain a logical possibility. The question now arises of whether such worlds (worlds of autotrophs plus plant-eating fauna only, that is) are compossible with the existence of the life-forms which we actually value, and, if not, whether they are remotely as valuable as worlds endowed with such life-forms. Rolston tackles these questions too, and answers as follows:

> An Earth with only herbivores and no omnivores or carnivores would be impoverished. The animal skills demanded would be only a fraction of those that have resulted in

---

95 G.J. Romanes ('Physicus') (1892), 'Supplementary Essay in Reply to a Recent Work on Theism', *A Candid Examination of Theism*, 152–80, p. 171.

96 Rolston, 'Disvalues', Section 2, 'Predation' (253–5), p. 253.

97 Rolston, ibid., p. 253.

98 Rolston, ibid., p. 253.

99 Brian Hebblethwaite (2000), *Evil, Suffering and Religion* (revised edn), pp. 112–13.

> actual zoology – no horns, no fleet-footed predators or prey, no fine-tuned eyesight and hearing, no quick neural capacity, no advanced brains.

And he proceeds to explain how predation enhances the capacities of prey species as well as of predators, and how the destinies of both kinds of species are intertwined.[100] Hence, if we quite generally value creatures with capacities resulting from predation, we have to value predation in general too. If, of course, we instead value nothing outside human life, then we have no such commitment (except where human predation is concerned). But resolutely anthropocentric stances such as this are rapidly becoming indefensible, and are therefore available neither to theists nor to their critics.

This approach, however, assumes that these capacities could not come about in any other way. Yet it must be logically possible for capacities such as fleetness of foot and quickness of sight and hearing to come about in the absence of evolution and natural selection. For example, a world such as that portrayed in the first chapter of Genesis is logically possible; and from that time onwards, species could have been providentially preserved in a fixed manner. But this would have required constant supernatural interventions in the natural order. In the absence of such interventions, characteristics such as reproductive capacities and competition for food were bound to govern the evolution of species. Thus, though evolution by natural selection is not logically necessary, it is probably the only kind of non-interventionist world-system which could give us those capacities found in nature that we value.

Just as Rolston does not claim that nothing could have been better, and that this is the best of all possible worlds (a notion discussed further in Chapter Seven, Section 1), I should not be construed as claiming that nothing could have been otherwise, and that the actual natural order is the only possible natural order. If a variety of single factors had been different, there would now be different species with relatedly different capacities. To select a large factor, numerous species which are as they are because Madagascar is an island would have been different if it had not been isolated from the land mass to which it used to belong; and this could plausibly have happened without the Big Bang or most of the laws of physics being much different. It cannot be held that, given natural selection but no continental drift, the species which now exist would have to be as they are. But, short of incessant supernatural interventions, the broad kinds of species that we have and that we value were not going to come into being in the absence of natural selection, and thus of predation.

Much that applies to predation applies to parasitism too, since parasitism (the ostensive target of Romanes's complaint) is a variety of predation. The additional problem is that parasites often forego the skills which their remote ancestors possessed, leading to what E.O. Wilson calls 'abject dependence on

---

[100] Rolston, ibid., p. 254.

the host species'.[101] However, as Rolston comments: 'The ill-adapted parasite species slays its host species and goes extinct; better-adapted parasites co-exist with their hosts, and the best-adapted will benefit their hosts at the species and even at the individual level.'[102]

Thus parasitism can and often does involve gains to the system of nature. Rolston still regards it as a disvalue, but adds that however much disvalue parasitism introduces, there must be a positive value in excess of it, else the parasite could not parasitize.[103] This comment, however, might be seen as inconclusive, since the issue is really whether nature might have had more value (or, to put things differently, have been better structured) in the complete absence of parasitism, rather than whether the segments of nature involving parasitism are of positive value overall. But Rolston has already argued that well-adapted parasites at least co-exist with the host without driving it to extinction, and often benefit the host, either as a species or as an individual, and this answers in part the issue just raised. Besides, once the intrinsic value of the lives of the parasites themselves (such as, say, cuckoos or mistletoe) is included in the reckoning, alongside their various impacts, good and bad, then, even if there are instances which have apparently more disvalue than value, a serious basis still turns out to exist for claiming that parasitism introduces more value than would exist in its absence. This is because the value of the life of parasitic creatures themselves could well exceed the cost to the creatures who are their victims.

Further, if the entire system involving natural selection has positive value overall, but (short of supernatural interventions) inevitably facilitates and involves parasitism, then the issue of the value or disvalue of parasitism is resolved, despite the losses to individual vegetable or animal victims. If so, the issue about the apparent overall evil of parasitism is to be regarded in much the same way as that of the apparent overall evil of predation. In terms of what we value, systems involving predation and parasitism are better than the possible alternatives. Yet as was argued in the previous section, when the value of worlds is under consideration, it is at the level of alternative law-governed systems that conclusions should be drawn and judgements formed; for a world governed by supernatural interventions would be no improvement. Predation and parasitism, then, form no significant problem for theists.

This chapter has concerned the consistency of Darwinism and theism, how Darwinism leaves intact the Teleological Argument, and how neither pre-Darwinian objections to theism and the created order (from Hume), nor parallel criticisms from Darwin, nor those from Darwinists who stress apparent evils such as predation and parasitism make their case. However, further aspects of the problem of evil, whether that of natural evil, as highlighted by Darwinism, or that of moral evil, remain for consideration. Issues such as the consistency

[101] Edward O. Wilson, op. cit., p.371; cited by Rolston, 'Disvalues', p. 276, n. 12.
[102] Rolston, ibid., p. 256.
[103] Rolston, ibid.

of evils with theism, and the problems distinctive to moral evil and to natural evil respectively, together with some of the interpretations of these matters from prominent Darwinists, are considered in Chapter Seven.

Chapter Seven

God and Evil

Section 1: Are Theism and Evil Compatible?

Some ways of reconciling a world of natural laws that include natural selection with the goodness of God have already come to light in the last two sections of the previous chapter. The system of nature has high value, and, as far as we have so far seen, could have been generated by a creator who favours such value; certainly no alternative system seems overall preferable. But further arguments against such a conclusion need to be considered, both arguments suggesting that belief in God's goodness is inconsistent with the facts of the world's evils, and others urging that even if there is no formal inconsistency, the scale and distribution of evils makes the world being the product of a good creator improbable. While the main focus will remain with evils like suffering and natural disasters that are not dependent on human decision or control (natural evil), it will be appropriate to consider also the bearing of moral evil, and whether the evils generated by humanity cohere with the world being God's world.

The general issue concerning consistency should first be introduced. In the terms of Stephen Davis, this is the 'Logical Problem of Evil'.[1] Many people have detected an inconsistency in holding that there exists an omnipotent, omniscient and good God and that evil exists. For an omnipotent and omniscient God would have the power and knowledge necessary to eliminate evil, and a good God would, it is suggested, have the inclination to do so. Since evil persists, there cannot be a God who is omnipotent, omniscient and good. Let us call this the stance of the atheologian.

However, it is difficult to establish an inconsistency, and many replies have also been advanced. Maybe a good creator would not eliminate evils either that are necessary for goods that would not exist otherwise, or the possibility of which is necessary for such goods to exist. Thus the possibility of human wrong-doing seems a requirement of the good of free human action, without which the world would be much less valuable. Indeed with this possibility there enters what has been called 'the Free Will Defence',[2] which argues that many evils are to be ascribed not to God but to human choice or free will; for a world with such freedom is more valuable than one without it, and the creation of such a world involves

---

1 Stephen T. Davis (1981), 'Free Will and Evil', in Davis (ed.), *Encountering Evil*, 69–83, p. 70.

2 Davis, ibid., p. 70.

the risk that wrong choices will be made resulting in evils. And where evils not generated by humanity are concerned, there could again be goods depending on natural systems which would not exist without these evils (like the flourishing of predators such as lions and eagles, dependent as it is on predation), as indicated in the last two sections of Chapter Six above.

Even if the truth of these replies is merely possible, the atheologian fails to establish the contradiction that he or she claims. For the propositions about God and about evil that he or she alleges to be inconsistent could after all be simultaneously true, and thus their conjunction need not be, as he or she implicitly claims, necessarily false. Indeed the suggestion that they involve a contradiction would be shown to fail if it is possibly true that a good creator would have reasons for not eliminating evil of which we are unaware. Unsatisfactory as this claim may seem, there is little prospect of it being known not to be possibly true.

All this depends, clearly, on there being a clear meaning for the terms in which the argument is couched, such as 'omnipotent', 'omniscient' and 'good'. Here is the definition of 'omnipotent', introduced in Chapter Four, Section 4 above:

> A being *x* is omnipotent if and only if it is capable of bringing about any contingent state of affairs (a) whose description does not contain or entail a contradiction and (b) whose description does not exclude or entail the exclusion of *x* or any omnipotent agent from among those which may have brought about that state of affairs.[3]

Omniscience was effectively defined in the previous paragraph of the same section. While this was done on the basis that God's knowledge (of whatever can be known) is timeless, those who adopt a temporal understanding of God's knowledge would be free to adapt that account to concern knowing whatever can be known at particular times. But they would have to cope with the problem of whether this includes necessary knowledge of the future choices of free creatures. (Anthony Kenny has argued that if it does include this, then determinism is true, and God is after all responsible for those choices.[4] But the timeless view seems to avoid this implication.)

As for 'good', this is an analogical term. The theory of analogy introduced above (see Chapter One, Section 6 above) suggests that sense can be made of such terms as long as something of the nature of a creator can be known, for in this way we can grasp the level and scope of God's goodness, on the analogy of human goodness (as understood against the level and scope of goodness of the human kind). If approaches of this type were to fail, then there would be no clear sense to talk of the divine goodness; at best, such talk would be merely metaphorical, as Kenny claims.[5] But all those who argue for or against the compatibility of God's

---

3 This is the definition of Bruce R. Reichenbach (1982), *Evil and a Good God*, p. 44.

4 Anthony Kenny (1979), *The God of the Philosophers*, p. 121.

5 Anthony Kenny (2004), *The Unknown God*, pp. 13–17, 34–45.

goodness and the facts of evil clearly assume that good, non-metaphorical sense can be made of such talk; for otherwise, God's goodness need have no particular connection with human goodness, and could be entirely distinctive, and there would be no discoverable problem of evil at all. While all these people could be wrong, I shall continue to rely here on the theory of analogy, and to assume that the problem is a genuine one. For if God has unlimited power and knowledge, and is a lover of value, the facts of evil do seem to present a problem, even if (as we have seen) not a contradiction.

That there is no contradiction could be shown by finding a possibly true proposition, consistent with one of the supposedly contradictory pair of claims, which, when conjoined with that claim, entails the other. In this connection, Davis puts forward the following: 'All the evil that exists in the world is due to the choices of free moral agents whom God created, and no other world that God could have created would have had a better balance of good over evil than the actual world will have.'[6] Now certainly this is consistent with 'God is omnipotent and God is good', and also possibly true, although the first part is so implausible (could a created devil really have caused all pre-human suffering?) as to make it a dubious proposition to select. But there is a worse problem; for it does not, strictly speaking, entail that evil exists (as opposed to entailing that any and every evil, if there are any, is due to the choices of free moral agents). It might also be criticized for suggesting that the actual world is the best possible world, given that this would not make sense, since there is no intrinsic maximum of goodness for worlds. But Davis could here defend his proposition, since there could be many possible worlds with an equally good balance of good over evil, as well as worlds with worse balances, and there might be none with a better balance, without someone committed to all this being committed to our world being the best possible world.

I suggest, then, that the problems be overcome by replacing Davis's suggestion with another, as follows: 'No other world that God could have created would have had a better balance of good over evil than the actual world, despite the many evils it contains, has or will have.' For this is possibly true, apparently consistent with God's omnipotence and goodness, and, in conjunction with God's omnipotence and goodness, entails that evil exists. At the same time it is not wildly implausible. Whether it is actually true will be discussed further, in later sections of this chapter; but it does not need to be known to be true to serve the purpose of showing that the claims about God and evil considered earlier do not comprise a contradiction.

An apparent contradiction could, however, be produced by the atheologian in response to the Free Will Defence. An omnipotent, omniscient and good creator, it could be held, would not create free agents who make evil choices, since such a creator could instead bring into being free agents who freely make none but good choices, and would have every reason to create just such free agents and no

---

[6] Davis, op. cit., p. 72.

others. But this conflicts with the fact of evil choices in the actual world. Since actual agents sometimes make evil choices, no such creator can exist.

Reasoning of this kind has been put forward by J.L. Mackie,[7] and discussed by Plantinga,[8] Davis[9] and Reichenbach.[10] While Plantinga, Davis and Reichenbach supply more detailed critiques than can be offered here, a reply to the atheologian could take the following general form. If God were to select for creation only creatures who make none but good choices, as opposed to creatures with genuine options open to them of doing good or evil, these would not be free creatures with respect to these so-called 'choices'. Indeed 'bringing it about that creatures freely choose to do only what is right' is itself a self-contradiction, and thus not within the powers of an omnipotent creator. Hence the apparent contradiction fails, and the Free Will Defence remains available to theists; for the good of genuinely free action is not to be had without the possibility that evil will be chosen. This means that it is also available when the argument concerning probability is under discussion, as it is in the coming section.

Section 2: Atrocities

While there seems to be no contradiction between theistic belief and the facts of evil, evils could still be too great or otherwise unacceptable for God's existence to be probable or for belief in God to be reasonable. This issue is regarded by Davis as beyond the scope of philosophy, as opposed to the realm of revelation and evangelism;[11] and yet many considerations relevant to the logical issue of consistency, recognized by Davis as philosophical ones, remain relevant here. Besides, philosophers often debate issues of probability, using Bayesian probability theory; indeed Reichenbach has devoted a chapter to applying this approach to the current issue, concluding that the atheologian's case has not been made.[12] Instead of discussing probability theory here, I will consider how far considerations already advanced are relevant to the problems of the amount and distribution of moral evil, problems such as atrocities and the maltreatment of the innocent. Corresponding issues relating to natural evil will be discussed in coming sections.

As Davis recognizes, one problem concerns the scale and amount of moral evil. Those theodicies (ways of reconciling God's goodness with the world's evils) that emphasize that pain and adversity are necessary for good qualities

---

7 J.L.Mackie (1971), 'Evil and Omnipotence', in Basil Mitchell (ed.), *The Philosophy of Religion*, 92–104; originally published in *Mind*, **64**, 1955, 200–12.

8 Alvin Plantinga (1971), 'The Free Will Defence', in Mitchell, ibid., 105–20; originally published in M. Black (ed.) (1965), *Philosophy in America*, 204–20.

9 Davis, op. cit., pp. 70–83.

10 Reichenbach, op. cit., pp. 64–86.

11 Davis, op. cit., pp. 75–7.

12 Reichenbach, op. cit., pp. 25–42.

like courage to be manifested are hard pressed to say how such atrocities as the Nazi holocaust are matched by corresponding goods; the scale of this atrocity could instead prompt the wish that humanity had not been equipped with such murderous powers.[13] Again, Fyodor Dostoevsky's character Ivan Karamazov recounts to his devout brother Alyosha tales of gratuitous cruelty towards the innocent, and challenges him to say whether he would agree to be the architect of a world in which one young child would be tortured to death as an inevitable aspect of a system designed to secure general peace and happiness. Alyosha replies that he would not.[14]

The issue here is not whether Alyosha would be willing personally to torture a child (much less whether God is to be regarded as doing this), but whether he would be willing to authorize a world in which there is a high risk that one or another free agent would do so, something which must be permitted if human history and culture are to be possible. The Free Will Defence ascribes direct responsibility for the evils chosen by free creatures to those creatures themselves, and not to God. Nevertheless, God has responsibility, according to theists, for creating a system enshrining the value of freedom and thus of the flourishing of free creatures, alongside the evils of suffering and death. The implicit suggestion of Ivan Karamazov is that it would be better not to create at all, since nothing can justify authorizing the torture and death of the innocent.[15]

But would it really be better not to create at all, so as to prevent unjust suffering and death? A partial analogy is the situation of parents, deciding whether to have children, who may lead worthwhile lives, may suffer unjustly, may lead worthwhile lives with some suffering, or may themselves perform acts of injustice. The risks are not usually considered to make having children wrong, although in imaginable circumstances things could be different. This is only a partial analogy, because of God's greater power and knowledge. Yet the suggestion (implicit in Ivan Karamazov's challenge to his brother) that God should deny the possibility of life and of flourishing to all free creatures because of the risk of injustice and atrocity seems a disproportionate judgement. Such a judgement suggests that disvalues either heavily outweigh or else trump positive values, in a manner that few if any human societies have ever accepted. It would be the equivalent for the realm of free and responsible creatures to the Thomas Hardy stance, considered in Chapter Six, Section 5 above, about the overall undesirability of sentient life.

A far more cogent stance is the evolutionary theodicy of John Hick, modelled on teaching of the second-century bishop Irenaeus (120–202). A good God, concerned for the mature development of free creatures, would furnish them neither with certainties, moral perfection nor with an ideal environment, but with ambivalent evidence, moral imperfection and an environment of challenge

---

[13] Davis, op. cit., p. 78.

[14] Fyodor Dostoevsky (1993), *The Brothers Karamazov*, Part II, Book V, Chapter 4; see also Brian Hebblethwaite (2000), *Evil, Suffering and Religion*, pp. 5–6.

[15] Hebblethwaite, pp. 67–70.

and danger; for the formation of mature character and of love of God through responses and choices with genuine alternatives, made in face of the genuine difficulties confronting early (and subsequent) generations of humanity, is of much greater value than guaranteed devotion.[16] While this theodicy is put forward in place of the Free Will Defence, which Hick rejects, these different theodicies are in fact compatible, since they differently account for God's permitting evil human choices (the Free Will Defence) and for relevant aspects of the human endowment and environment (the Irenaean theodicy). Neither theodicy can cope with the full extent of natural evil, despite the hopes of Davis and Hick respectively. But Hick's discussion of human origins and of the development of early human culture valuably adds to other theodicies in a manner that also takes seriously evolution and the struggle for survival, as well as the moral and spiritual capacities of humanity.[17] While Hick's full case depends in part on God's purposes being fulfilled in a realm situated after death, those unable to subscribe to this premise remain free to adopt his theodicy about past, present and, perhaps, future generations and the conditions of human life that a good God might generate.

Yet the atheologian has another resort, relevant in particular to large-scale atrocities such as the holocaust, as well as to more isolated injustices and acts of gratuitous cruelty. Would not a good God intervene to prevent the worst moral evils facilitated by the system of natural laws and of creaturely freedom? Thus Hume's suggestion that a few touches to Caligula's brain would have produced a better world could be modified to concern intervention affecting the leaders of the Third Reich. The absence of intervention in such cases might thus be treated as evidence for the improbability of theistic belief.

However, the atheologian's case here assumes that a world involving miraculous interventions (few or many) to prevent great evils would be better than a regular world without such interventions. As was noted when Hume's suggestions for a better creation were discussed, a world in which (some or all) impending calamities failed to take place (with sinking ships and collapsing towers unexpectedly righting themselves) would be a world unamenable to scientific understanding and prediction, and one in which the consequences of action were liable mysteriously and inexplicably to disappear or be counteracted. (Nor will it do to suggest that scientific laws never apply universally, and are thus compatible, one and all, with such interventions. For even when apparent counter-examples to current universal formulations come to light, we assume that there are underlying universal regularities and that the counter-examples can be subsumed within some more comprehensive and universal formulation. Probabilistic laws, certainly, admit of counter-examples, but not just any counter-examples whatever; they preclude, for example, such macroscopic effects as the resurrection of collapsing towers and the

---

16 John Hick (1966), *Evil and the God of Love*; Hick, 'An Irenaean Theodicy', in Davis, op. cit., 39–52.

17 Hick, 'An Irenaean Theodicy', pp. 45–7.

reinstatement of suicides to the precipices from which they have leapt.) Indeed in a world in which the regular course of events and the natural consequences of action sometimes inexplicably failed to unfold, the entire context of human life and action would be unrecognizably different. There is no need to elaborate here a value-theory by which possible worlds could be compared; for it is in any case far from obvious, whatever one's value-theory may be, that the kind of world currently suggested would be an improvement. I am not suggesting that losses to freedom outweigh everything else. For the suggested world-order would strike not only at free choice but also at the possibilities for rationality, and at the overall context for the formation of character and relationships, or, in other words, at the development of persons. (It would of course strike at the evolution of species too.)

The argument against the probability of theistic belief needs to claim that the actual world order is not as likely or as probable to be the creation of a good God as some other that could have been created, or as no creation at all. But where moral evil is concerned, no consistent better alternative system has been depicted. (If the concept of probability is inapplicable to such issues, a view that has been discussed but rejected in Chapter Five, Section 5 above, then the atheologian is in difficulties for this reason too. But the atheologian's case could, if necessary, be re-expressed in terms of whether the actual world order is more credible as the creation of a good God than another or than none, in which case the above replies could be repeated with appropriate adjustments.) It remains to be seen, in the coming sections, whether or not the same holds where natural evil is concerned.

Section 3: Natural Evil

Even if theistic belief is not undermined by moral evil, there remains the issue of whether it is not rendered improbable (or less than credible) by natural evil, its extent and its distribution. Maybe no formal inconsistency is in question, since a good God could have reasons for permitting natural evils as well as moral evils. But the extent and distribution of natural evils still generate such problems for the believer as to comprise the core of the atheologian's case against the tenability of theism. This case, as we have seen, was implicit in several of Hume's suggestions for a better world, and in Darwin's doubts about whether a good God could create the Ichneumonidae. More recently it has been restated in the work of Richard Dawkins.[18] We have also encountered theistic responses from Rolston to the problems of predation and parasitism, responses relating to the natural laws and systems involved. Before addressing Dawkins's presentation of the issue, we should first consider whether similar responses are available for the evils of

[18] Richard Dawkins (1995), *River Out of Eden*, pp. 111–55.

pain and suffering, as offered by Reichenbach and Rolston when they discuss alternatives to the world as it is.

Hume's suggestion that evils could be avoided if the world were less regular has already been seen (in Chapter Six, Section 4 above) to have unacceptable implications. A world of direct supernatural governance, lacking natural regularities, would be unpredictable and incomprehensible, and the lives of creatures would be unrecognizably different both from current reality and from the kind of lives Hume would have wished to preserve; and much the same applies to Hume's second scenario, in which laws of nature operate but God supersedes them to prevent the occurrence of evils. Reichenbach adds that 'a world run by divine miraculous intervention would be incompatible with a world inhabited by significantly free agents'. For agents' brain-states would be liable to be modified to prevent evil, and in any case decisions to perform evil deeds could not be implemented. Such a world would not be selected by a creator seeking to actualize 'the greater good',[19] at least if we accept Reichenbach's presupposition that: 'A world containing significantly free persons making choices between moral good and evil and choosing a significant amount of moral good is superior to a world lacking significantly free persons and moral good and evil.'[20] Even if we prefer to qualify this presupposition, possibly on the grounds that it takes animal wellbeing and suffering insufficiently into account, the qualified version is still likely to recognize the superiority of a world containing significantly free persons, other things being equal, and for most circumstances this gives Reichenbach what he requires, and allows him to draw his conclusion.

Accordingly, as Reichenbach goes on to conclude, 'for God to actualise the greater good it was necessary that the world he create operate according to natural law'.[21] But since 'natural evils are a consequence of natural objects acting according to natural laws upon sentient, natural creatures', God cannot be held blameworthy for them, since actualizing the greater good requires the actualizing of all this.[22] (To cover the possibility of natural evils befalling non-sentient living creatures, Reichenbach's claim could be adjusted and broadened so that 'living' replaces 'sentient'.) He does, however, recognize and discuss the possibilities that God could employ different natural laws, or create a world occupied by different entities, such as one lacking poisons or parasites.[23]

However, as Reichenbach proceeds to argue, altering natural laws to eliminate evils involves altering the constituents of the world. Changing combustion so that creatures would not be burned involves making them non-combustible, and changing fire so that it does not burn them. Preventing the capacity of water to drown us would mean changing its specific gravity, and many of its beneficial

---

19 Reichenbach, op. cit., p. 105.
20 Reichenbach, ibid., p. 47.
21 Reichenbach, ibid., p. 106.
22 Reichenbach, ibid., pp. 106–7.
23 Reichenbach, ibid., p. 110.

properties in consequence, including, as F.R. Tennant mentions, 'its thirst-quenching and cleansing functions'.[24] Averting evils befalling humanity, for many of which sentience is pivotal, could even involve changing humanity so as to discontinue human sentience.[25]

But these counter-intuitive implications stem from attempts to describe a world without natural evils. Other possibilities should also be considered in which sentience remains but natural evils are reduced. One example would be a world lacking pain-producing micro-organisms.[26] This possibility was effectively considered above in Chapter Six in connection with parasitism; it would seem to require either a cataclysmic intervention, or a world without natural selection and thus without the complexity and adaptation of the living creatures of our world. Others might involve changing laws of nature, for example, whatever natural laws facilitate cancer. But as Reichenbach remarks, and as I have argued elsewhere, such would-be descriptions of a better world are unsatisfactory unless the overall impacts of the proposed differences are taken into account and found to involve an improvement.[27] In the case of cancer, different laws of cellular growth and multiplication could well prove incompatible with the normal growth and development of creatures and of their embryos; for better laws would have to sustain processes recognized to be good and healthy in the world as it is, as well as reducing specific evils that we may deplore.

What the atheologian needs to supply is a consistent description of a better natural system (with either different natural laws, different components, or both), and to show that this system would result in a better balance of good over evil than that of the current world-system. But, as Reichenbach adds, this is a task that cannot easily be accomplished, and is suited to nothing less than an omniscient mind.[28] Yet, until it is accomplished, we have no reason to believe that a world with a better balance of good over evil than the actual world is possible, or that the actual world is not a world that a good God would create. For the value that the atheologian seeks to preserve is apparently not to be had without the kind of evils that the atheologian seeks to eliminate.

A good example of this is the problem of waste. For myriads of young lives are generated in nature, from spores and seeds to fledglings and foetuses, most of which die long before they can develop into mature creatures; and this is widely held to be a problem for belief in a good God, whether because resources are supposedly squandered, or because of the lost potential of aborted creatures. Yet

---

[24] F.R. Tennant (1928–30), *Philosophical Theology*, vol. II, p. 201.

[25] Reichenbach, op. cit., pp. 111–12.

[26] Reichenbach, ibid., pp. 114–15.

[27] Reichenbach, ibid., pp. 115–16; Robin Attfield (1993), *God and The Secular* (2nd edn), pp. 206–9.

[28] Reichenbach, op. cit., p. 116.

Rolston is right to remark that there is little or no waste within organic nature,[29] or rather there was little or none before the human invention of non-biodegradable substances such as most plastics. As Rolston says, it is 'this alleged "waste"' which 'makes trophic pyramids possible. Wherever there is available free energy and biomass, a life form typically evolves to exploit these resources'.[30] Apparent waste, then, is (as Rolston suggests) a characteristic of systems that make the continuing evolution of complex life a possibility. Indeed, much the same can be said about another form of natural waste, the loss of species in the course of evolutionary history (as opposed to species-loss caused by humanity). This loss of species is sometimes considered a theological problem.[31] But if the evolution of complex life is only to be had through natural selection, then because natural selection inevitably involves the elimination of large numbers of species, species-loss turns out to be an indispensable feature of any system that makes such life possible.

Rolston's treatment of disvalues in nature focuses more specifically on pain and suffering (often considered to be what makes waste a problem), and expounds further systemic features of the actual world-system. It also refers back to his discussion of predation (discussed in Chapter Six, Section 5 above). The sensory and cognitive capacities and skills which we value in nature have only been made possible in the context of predation, both for predator and for prey, and suffering is an indispensable aspect of this process, without which these skills and capacities would not have developed. The capacity to suffer drives the capacity for focussed consciousness; thought cannot happen in plants, and emerges in the course of evolutionary history partly so as to secure wellbeing while at the same time averting pain; for this seems to be one of its central evolutionary functions. In general, natural selection requires pain as an alarm system, promoting extra agility, intelligence and acuteness of perception; it also selects in favour of functional pain and against pain which is counterproductive.[32] Thus without suffering (and in general without natural evil) much of what we value in the natural world, and also in human life, would also be absent.

Rolston's case is impressive, without being conclusive. Certainly, if predation is to be accepted as an aspect of an overall valuable system, so must pain and suffering be. But given large numbers of painful deaths and many cases of intense non-functional pain over long tracts of history and pre-history, the verdict might linger that the system could still be wanting in terms of value. What should be added to the balance of the argument here is the positive value of the flourishing of each creature that has ever lived and flourished, including those of pre-human

---

[29] Holmes Rolston III (1992), 'Disvalues in Nature', *The Monist* **75**, 250–78, pp. 268–70.

[30] Rolston, ibid., p.269.

[31] Ruth Page (2004), 'Panentheism and Pansyntheism: God in Relation', in Philip Clayton and Arthur Peacocke (eds), *In Whom We Live and Move and Have Our Being*, 222–32.

[32] Rolston, ibid., pp. 272–3.

times. If it is granted that the flourishing of members both of prey and of predator species is positively and independently valuable[33] (and thus not valuable only in a derivative or instrumental sense), then the value of the flourishing of these organisms will not have to depend, as some theodicies (Irenaean and otherwise) mistakenly claim, solely on the *human* capacities which it indirectly facilitates through evolutionary processes, and on the value of personal development thus facilitated. Further, the flourishing of the animals concerned depends, as Rolston makes clear, on predation and on the capacity for suffering which pervades the realms of conscious animal life. Thus a great deal of what we value depends for its existence on suffering (past and present), as well as on other aspects of the great system of nature. There could have been another system, for example one lacking sentience or suffering, but very little of what we value could be included in it.

Seen in this light, the system of nature may well be regarded as having an overall balance of value over disvalue. There is widespread disvalue, immense evil, within it, including suffering, disease, injury and early death. But these types of evils are systemic preconditions of the flourishing of billions of creatures across the ages, as well as of human capacities and of the human endowment. While the coexistence of widespread flourishing and the absence of evils of this extent would be logically possible, yet short of equally widespread supernatural intervention taking place with a frequency that would make nature irregular and fundamentally unpredictable, and life unrecognizable, the only alternatives to evils of this extent are a world with no life (and hence no value) at all, and a world with non-sentient life only, with the limited kinds of predation and the limited kinds of value which such a world could sustain. Thus none of the possible alternative scenarios seems remotely to compare in value to the actual system of nature.

Section 4: Dawkins on Predation

Richard Dawkins advances an apparently formidable argument from the phenomena of predation and parasitism against belief in a good creator. After discussing the parasitism of the Ichneumon wasp, which ensures that its victim is paralyzed but kept alive, he asks what cheetahs and the antelopes who are the cheetahs' would-be prey are designed for, calling this 'God's utility function'. His conclusions are that they are not designed for anything but DNA survival, that there is no divine engineer, and that nature is indifferent to the good or the happiness of individual creatures.[34] The suggestion is that if there were a creator,

---

33 Or intrinsically valuable. But it is convenient here to defer the explicit introduction of this notion until it can be clarified and contrasted with a neighbouring concept in Chapter Eight, Section 1 below. It is also discussed in Chapter 3 of Robin Attfield (1995), *Value, Obligation and Meta-Ethics*, and in Attfield (2001), 'Postmodernism, Value and Objectivity', *Environmental Values*, **10**.2, 145–62.

34 Dawkins, *River Out of Eden*, pp. 111–55.

nature would operate differently. To this argument, Michael Ruse has presented an ingenious reply on behalf of theistic believers, based on aspects of Dawkins's adaptationism.[35] Dawkins's argument, Ruse's reply, and relevant variations on them (relating to alternative versions of Darwinism) will be discussed in this section and the next.

Dawkins is a strongly adaptationist Darwinian, who holds that nothing but natural selection can explain the complex adaptation of living creatures. Neither Lamarckianism (with its principles of development through use and of the inheritability of acquired characteristics) nor mutationism (with its belief in hopeful monsters bridging apparent evolutionary gaps) can explain this;[36] and nor can Paley-like belief in a watch-maker designer, which Darwinism shows to be redundant.[37] This is partly because faith 'means blind trust', adopted 'in the absence of evidence',[38] an account of theistic belief soundly confuted by his critic Alister McGrath.[39] And it is partly because, as we have seen in Chapter Five, Section 4 above, Dawkins thinks that a divine designer would be so complex as to be inexplicable, since what is complex requires an even more complex explanation.[40] This reasoning rests, as Ward and Ruse have both remarked, on an unreliable premise about explanations, since complex and apparently disconnected phenomena sometimes prove to have simple explanations, and also on the unreliable presuppositions that a divine designer would have to be complex (and not unified) in much the same sense as the complex, disconnected thoughts of a human mind, and that the existence of such a designer would not in any case explain anything unless it were explicable by something else.[41] Hence Dawkins's attempts to derive atheism from Darwinism are misguided. So, arguably, is his tendency to treat religious belief as irrational. Yet his argument from predation could still have independent merit.

In discussing cheetahs and antelopes, Dawkins argues that each of them appears well designed, cheetahs to kill antelopes, and antelopes for the survival of antelopes and starvation among cheetahs. It is almost as if there were two opposing deities at work. If just one creator is responsible, 'what is he playing at? Is he a sadist who enjoys spectator blood sports? Is he trying to avoid overpopulation in the mammals of Africa? Is he manoeuvring to maximize David Attenborough's television ratings?' The true answer is supposedly that there is no true or cogent

---

35 Michael Ruse (2001), *Can a Darwinian Be a Christian?*, pp. 136–37; Ruse (2003), *Darwin and Design*, pp. 330–33.

36 Dawkins (1988), *The Blind Watchmaker*, pp. 287–318.

37 Dawkins, ibid., pp. 4–37.

38 Dawkins (1989), *The Selfish Gene* (2nd edn), p. 198.

39 Alister McGrath (2005), *Dawkins' God*, pp. 84–91.

40 Dawkins, *The Blind Watchmaker*, 141.

41 Ward (1996), *God, Chance and Necessity*, pp. 110–12; Ruse, *Can a Darwinian?*, pp. 114–15. See also McGrath, op. cit., pp. 90–91. The second presupposition here is discussed and rejected in Chapter Five, Section 4 above.

answer to these questions, for there is no 'utility function' but the survival of cheetah and antelope DNA.[42] Granted Darwinism, then, the phenomena of life seem to suggest that there cannot be a good or benign designer.

Ruse remarks at this point that theistic believers are here prone to reject Darwinism and adopt a less unfriendly form of evolutionism. The philosopher he cites in this connection is Holmes Rolston, whose stress on the self-organizing powers of creatures and belief in upward progress within evolution Ruse regards as non-Darwinian.[43] An assessment of this appraisal can be deferred to the coming section. Here I need only remark that when Rolston once quoted and discussed Ruse's words: 'The secret of the organic world is evolution caused by natural selection working on small, undirected variation',[44] he should probably not be construed as rejecting them, as Ruse seems to suppose, but as seeking to supplement them (in ways to be returned to below).[45] Importantly, Rolston's discussion of 'Disvalues in Nature', considered in Chapter Six, Section 5 above and in the previous section of the current chapter, is fully in keeping with Darwinian accounts of predation, parasitism and suffering.[46] There is no need either to retract conclusions there drawn, or to gloss them as un-Darwinian.

The suggestion that Rolston diverges from Darwinism could also be directed at Ward when he seeks to supplement Dawkinsian, gene-machine-like, accounts of evolution by reference to a guiding purpose.[47] Discussion of these themes too can be deferred (to Chapter Eight), but it is relevant here to remark that when Ward points out that the utility-function of cheetahs and antelopes is not just DNA survival but, in the first instance at least, 'the gradual rearrangement of DNA codes, to form ever-more sophisticated programs for building bodies',[48] and that the point of this could lie in the bodies and the related lives rather than in the programmes, he is both correcting Dawkins and working within Dawkins's own Darwinian principles, while leaving open the possibility that the existence of these bodies, and whatever values they may carry, is another, perhaps ulterior, part of the utility-function (or purpose).

Ruse, however, prefers to stay within the Darwinism of Dawkins's variety, and produces on behalf of the believer a powerful refutation of Dawkins's argument. For according to Dawkins, adaptive complexity never comes about except through natural selection. (There can be other forms of evolution, but not ones leading to adaptive complexity.) Ruse takes Dawkins to be saying that without natural selection, adaptive complexity is impossible. But God cannot do the impossible. So the only options open to the creator were (a) not to create at all, or, if the

---

42 Dawkins, *River Out of Eden*, p. 123.

43 Ruse, *Darwin and Design*, pp. 308–11, 330.

44 Holmes Rolston (1987), *Science and Religion*, p. 91.

45 Ruse, *Darwin and Design*, p. 309.

46 Rolstons (1992), 'Disvalues in Nature'.

47 Ward, op. cit., pp. 136–41; see also p. 72.

48 Ward, ibid., p. 137.

existence of life-forms in all their variety were the goal (or part of it), then (b) to create a world of natural selection, complete with predation, parasitism, agony and suffering and (apparent) waste. If the existence of actual, existing life-forms was among God's values or goals, then there was no other way; a good creator would create a world of natural selection after all. Indeed Darwinism 'may even make the solution' to the problem of natural evil 'easier to grasp'.[49]

If Ruse is interpreting Dawkins correctly, and if Dawkins's version of Darwinism is accepted, then this theodicy (which Ruse rightly regards as a development of Reichenbach's) seems successful. So a key issue is whether Dawkins is really saying that adaptive complexity is impossible without natural selection. The evidence of the final chapter of *The Blind Watchmaker* and of Dawkins's essay 'Universal Darwinism' seems to support this interpretation. Dawkins holds not just that no process other than natural selection brings about the adaptive complexity of life;[50] he is also, he relates, prepared to bet that life will always be Darwinian life, and that the 'Darwinian Law ... may be as universal as the great laws of physics'.[51] He even asserts that 'if a form of life is ever discovered in another part of the universe, ... it will be found to resemble life on Earth in one key respect: it will have evolved by some kind of Darwinian natural selection'.[52] It is from passages such as these that Ruse draws his impossibility conclusion.

Ruse is most obviously justified in ascribing to Dawkins the view that whatever could befall in the actual world would accord with the Darwinian Law, and this already supports counterfactuals about what could or could not have happened in the actual world. But the laws of physics, to which Dawkins compares the Darwinian Law, do not have all possible worlds as their scope, and yet Ruse's argument requires Dawkins to hold that life without natural selection does not occur in any possible world whatever. (Otherwise an objector to Ruse could argue that a world with life but without natural selection is possible, and so an omnipotent creator could create it after all.) Ruse here seems to leap from what is physically possible in the actual world to what is metaphysically possible; and it is not obvious that Dawkins had metaphysical possibility or impossibility in mind. Yet Dawkins would clearly wish to adhere to the applicability of Darwinism to any possible world at all similar to the actual world, for example ones with slightly more or slightly fewer stars, but still suited to life. Besides, Dawkins's reasons for regarding Darwinism as universal seem to apply across worlds; thus large or sudden mutations would not, in his view, produce adaptive complexity in any world-system. Thus, despite the shift of modality (from physical to metaphysical impossibility) mentioned above, Ruse's interpretation of Dawkins is probably a reasonable one. Given also the truth of this kind of Darwinism, Ruse's theodicy

---

49 Ruse, *Darwin and Design*, p. 333.

50 Dawkins, *The Blind Watchmaker*, pp. 287–318.

51 Dawkins (1983), 'Universal Darwinism', in D.S. Bendall (ed.), *Evolution from Molecules to Men*, 403–25, p. 423.

52 Dawkins, *The Blind Watchmaker*, p. 288.

would be successful, and Dawkins's argument from predation would be effectively countered.

But the possibility should also be considered that in some possible world there could be life without natural selection. Just possibly Dawkins is not denying this for all possible worlds after all, or, even if he is making such a strong denial, he could be wrong. Or possibly evolution can take place with natural selection as just one mechanism among others, albeit the key one, rather as Gould and others suggest. There again, possibly the kind of Darwinism supported by Dawkins needs to be supplemented, as suggested by Rolston and Ward. These possibilities and their bearing on the problem of suffering are reviewed in the coming section.

Section 5: Could Things be Otherwise?

First, we need to consider whether there could be life (complex or otherwise) without natural selection, in case Dawkins (as interpreted by Ruse) is wrong, and Ruse's theodicy (see the previous section) needs to be revised. Is this possible? Since the existence of God and the creation of a world run by miracles are both possibilities, rather as creationists believe about the actual world, the answer has to be 'yes', despite the denials of strong adaptationists. But on the assumption that the actual world operates by natural selection, the mere possibility of life being generated by miracles without natural selection in worlds that are not actual does not resuscitate problems for theodicy; for, as has been argued above, a world run by miracles would not be an improvement on the actual world. Theistic Darwinists such as Fisher have believed that God knowingly forwent this option to create a world with a better balance of good over evil.

However, the possibility of non-miraculous worlds harbouring life but with natural selection as just one mechanism among others, albeit the key one, should also be investigated. As has been mentioned, this theory has been seriously suggested for the actual world by Gould and Lewontin in their spandrels paper.[53] Dawkins considers this theory impossible, since nothing but small adaptations can generate biological complexity,[54] but he also recognizes that Darwinism in general may one day have to be discarded.[55] In these circumstances, the Gould/Lewontin stance cannot be rejected as epistemically impossible, for the actual world as well as possible ones.

But a version of Ruse's reply to Dawkins will once again be relevant. For if natural selection remains indispensable for the development of complex life, then a God who desires the existence of complex life will create a world in which

---

[53] Stephen Jay Gould and Richard C. Lewontin (1994), 'The Spandrels of San Marco and the Panglossian Paradigm', in Elliott Sober (ed.), *Conceptual Issues in Evolutionary Biology* (2nd edn), 73–90.

[54] Dawkins, *The Blind Watchmaker*, p. 288.

[55] Dawkins (2003), *A Devil's Chaplain*, p. 81.

natural selection plays this indispensable role; but this role will be one among others, supplemented, that is, by other processes. It can be added here that if the role of natural selection is less universal or comprehensive than in Dawkins's theory, then the actual world will have seen somewhat less predation, parasitism and consequent suffering than Dawkins's theory retrodicts, although the extent of each will still have been vast. For the creator will have actualized somewhat less predation, parasitism and related suffering, than would have been created if Dawkins's theory of what is possible had been the full story. Hence the problems for theodicy are slightly less acute if this version of Darwinism is correct.

Very similar comments apply to the theory of Lewontin that natural selection does not invariably involve competition for resources, and thus predation or parasitism, and can instead operate through opportunism, inventiveness and differential rates of reproduction. For the same reasons, this theory cannot be ruled out as possibly true of the actual world. It does not, however, deny the indispensability of natural selection. Hence Ruse's reply to Dawkins still holds; if natural selection is indispensable for complex life, a creator desirous of complex life will actualize a world of natural selection. But this time, natural selection itself involves less predation, parasitism and suffering than it involves on Dawkins's theory, and so once again the problems for theodicy are less acute than they might seem. Parallel remarks will apply to other partially non-adaptationist versions of Darwinism insofar as they are compatible with natural selection (such as theories blending adaptationism with a little mutationism).

The possibility should, however, be revisited of life coming about without miracles but without the suffering associated with natural selection either, or at least without the kind of predation conducted by carnivores (including the Ichneumon wasp). Here we are back with possible worlds; but if such a world is possible, it would contain, at most, vegetables and unicellular organisms, and, as Rolston argues, there is no reason to suppose that it would contain consciousness or that consciousness could there evolve. It does not, in my view, follow that it would be valueless. Indeed a thought-experiment once conducted by Donald Scherer concludes that most people would recognize in such a world one of much greater value than a world entirely lacking in life.[56] But in view of the considerable value that consciousness facilitates, it does not compare in value with worlds such as ours, and no one is likely to suggest that a good creator would have preferred such a world to the actual one.

But does Darwinism need to be supplemented as suggested by Rolston and later by Ward? Ruse, for example, presents Rolston's stance as, potentially, a different kind of response to the Darwinism of Dawkins from his own.[57] It should be stressed, however, that Rolston is not saying that Darwinism should be replaced. To quote his words: 'natural selection is only a fraction of some bigger truth. We

---

56 Donald Scherer (1983), 'Anthropocentrism, Atomism and Environmental Ethics', in Donald Scherer and Thomas Attig (eds), *Ethics and the Environment*, 54–81.

57 Ruse, *Darwin and Design*, pp. 308–11.

don't want Darwinian theory cast down, but we do want it recast.'[58] For unlike Dawkins[59] and Ruse,[60] Rolston considers that Darwinism fails to explain the appearance of the full range of adaptive complexity. He grants that it shows how adaptations are possible, but considers that it does not explain overall change in the direction of complex life-forms, when change in the other direction would be equally explicable. While there is a slight apparent overlap between this approach and that of creationists, Rolston is not invoking supernatural intervention or a 'god of the gaps', and rather sees himself as developing Gould's belief that there may be explanations additional to natural selection.[61] For what he has in mind is a naturalistic supplement capable of explaining 'the push-up, lock-up effect by which inorganic energy input … can spontaneously happen to synthesize negentropic amino acid subunits … which would be degraded by entropy, except that … they make themselves resistant to degradation'.[62] (Similarly the distinguished scientist Arthur Peacocke writes of natural selection being 'perhaps supported by some other factors facilitating complexification'.[63]) What is depicted here is a type of theory, not a specific theory, and one essentially complying with 'naturalistic pathways'.[64] Thus there is no specific theory here to appraise; but we should not rule out the possibility of such a theory beforehand, given that it is not committed to supernatural explanations as supplements of naturalistic ones.

Possibly the same should be said of Ward's advocacy of explanations additional to Darwinism.[65] It may seem less obvious than with Rolston and Peacocke that Ward has naturalistic explanations in mind when he adds to Darwinism talk of a guiding purpose,[66] but Ward has no wish to invoke supernatural interventions. His point is rather that, just as the emergence of life depends on the creation of the right kind of laws and initial conditions, so does the occurrence of the right range of mutations, of the right sets of environmental conditions, and thus of the process of natural selection, if adaptive complexity is to ensue.[67] Thus, for Ward, creation by God (as understood by theists, and as expounded and defended in chapters Four and Five above) forms a key part of the explanation of evolution. This account appears to presuppose a naturalistic supplement to natural selection, of the 'push-up, lock-up effect' kind depicted

58 Rolston, *Science and Religion*, p. 121.

59 Dawkins (1996), *Climbing Mount Improbable*.

60 Ruse, *Can a Darwinian?*, pp. 115–22.

61 Rolston, op. cit., p. 120.

62 Rolston, ibid., pp. 111–12.

63 Arthur Peacocke (2004), 'Articulating God's Presence in and to the World Unveiled by the Sciences', in Philip Clayton and Arthur Peacocke (eds), *In Whom We Live*, 137–54, p. 141.

64 Rolston, op. cit., p. 123; Peacocke., op. cit., p. 142.

65 Ward, *God, Chance and Necessity*, pp. 136–41.

66 Ward, ibid., pp. 61–95.

67 Ward, ibid., p. 68.

by Rolston, but grounds the very capacity of natural selection to explain the emergence of biological complexity in the original creation and thus in divine purposes. These purposes, through their embodiment in the right laws and initial conditions, are regarded as guiding the entire process. This theory has the further advantage for the theist that the randomness of random mutations is regarded not as irreducibly serendipitous and dysfunctional but as an aspect of an underlying plan; as D.J. Bartholomew has written, 'there is every reason to suppose that a Creator wishing to achieve certain ends might choose to reach them by introducing random processes whose macro-behaviour would have the desired character', where the ends might be (or include) 'to produce intelligent beings capable of interaction with their Creator'.[68] (Ward's theory of value, introduced in reply to Dawkins to cast light on what may really be 'God's utility function', is discussed in Chapter Eight below.)

If Darwinism needs to be supplemented by supplementary naturalistic explanations, as these writers suggest, then Ruse's response to Dawkins requires adjustment, but not replacement. For if natural selection is indispensable as an explanation of adaptive complexity (even if it is insufficient), then God could not create complex life without it, and if desirous of such complexity would actualize a world of natural selection. Additional naturalistic factors would be generated or facilitated too, if adaptive complexity could not be had without them; until and unless a defensible para-Darwinian theory is produced, we cannot tell which these would be. Overall, the conclusion emerges that whichever form of Darwinism is accepted, versions of Ruse's response continue to stand up, and that this applies also to supplementary theories, or to what might be called Darwinism-plus. In no case are the facts of natural evil (any more than the facts of moral evil) incompatible with belief in a good creator, or prone to render that belief improbable or unacceptable. This conclusion stands whatever view is taken about the possibility of life after death, sometimes claimed in theodicies to contribute to the grounds for this belief;[69] for this possibility lies beyond the scope of the present book.

---

68 D.J. Bartholomew (1984), *God of Chance*, p. 98.

69 Thus John Hick (1963), *Philosophy of Religion*, pp. 40–47, 50–53.

Chapter Eight

Purpose, Immanence and the Argument from Value

Section 1: Ward's Value Argument

There is more to be said about the value present in nature and relatedly about the purpose that may underlie evolution. To both these topics Keith Ward has made important recent contributions, from which much can be learned. His argument from value will first be introduced, and then criticized. This will allow a refined version of the argument from value to be developed, and related arguments in the literature to be traced. It also facilitates a related discussion of whether evolution is progressive, and might have a goal. This discussion allows the topics of God's immanence and of the relation of Darwinism to theism to be reappraised. The obvious starting-point is Keith Ward's argument from value.

This is a new kind of design argument, according to which design on the part of a purposive and loving creator is a reasonable interpretation of the immense amount of intrinsic value in the world as we find it. The world of nature is strikingly full of valuable states of affairs, and a rational interpretation of this phenomenon consists in its being the creation of a creator desirous of such value.[1] (Ward construes this argument as neither deductive nor inductive, but rather as comparable to interpretative arguments in music or art.) This promising argument turns on (and, I shall argue, is weakened by) Ward's adoption of a definition of intrinsic value which makes a state of affairs have intrinsic value as long as one has reason to value it *whatever* one values. This definition clearly makes knowledge, power and wisdom into intrinsic values, and just as clearly makes the wellbeing of non-human creatures not to have value of this kind. Other states recognized by Ward as intrinsically valuable include awareness or consciousness, pleasure, creativity, self-determination and free activity,[2] plus co-operation in the pursuit of freely chosen purposes.[3]

Ward has subsequently developed his argument from value, relating it to a progressive interpretation of evolution. Because the world's natural processes have so evolved as to provide for the emergence of consciousness and other intrinsically desirable states such as creativity and collaboration, and since these are just the

1 Keith Ward (1982), *Rational Theology and the Creativity of God*, 89–120.

2 Ward, ibid., p. 104.

3 Ward, ibid., pp. 105–6.

states that a loving creator, desirous of what is intrinsically desirable, could be expected to seek to generate, there is good reason to regard evolution as having such desirable, complex states as an end, and as being chosen to supply that mix of regularity and freedom for finite creatures that make the emergence of these desirable end-states possible.[4] Thus evolution is to be regarded as progressive and purposive, despite the denials of Darwinists such as Ruse,[5] and intrinsic values, here defined as 'states that all sentient beings have reason to want' (and not the survival of units of DNA), ultimately comprise (in Dawkins's phrase) God's 'utility function' or purpose.[6] Hence mind would not be a chance by-product of evolution, as the gene-machine version of Darwinism appears to suggest, but (at least) part of its underlying rationale.[7]

More recently, in an essay entitled 'Theistic Evolution', Ward contrasts this creativity-oriented view with an imaginary world in which God determines every specific event for the best, and makes no provision for 'the causality of finite agents', or for 'libertarian freedom', 'that property of a rational agent by which, on at least some occasions, no prior physical state, even with the addition of a set of general laws of nature, entails one specific outcome of a given situation',[8] and through which the outcome is determined in part by that agent. If consciousness and those valuable states like the appreciation of beauty, reflective thought and the quest for truth that it facilitates are among God's purposes, then the creator would provide for their emergence through natural processes, and thus embody indeterminacy (arising, for example, through natural selection among mutations) in the processes that facilitate the nature of prospectively conscious creatures. For indeterminacy is a necessary condition (though not a sufficient one) of libertarian freedom.[9] (Bartholomew's argument about how stochastic mutational processes could well be employed for the sake of purposive outcomes (mentioned above in Chapter Seven, Section 5) would again be relevant here.)

Thus Ward applies his argument from value on a broad front; but we need to return to his definitions of intrinsic value. His earlier definition (what one has reason to value whatever one values) accords intrinsic value to items normally regarded as having instrumental value, such as power, rather than having intrinsic value (in the usual sense of the phrase), and is also mainly if not exclusively concerned, or so his examples suggest, with goods instrumental to human purposes rather than to those of other creatures. While it is true that the world is replete with such goods, and could have been less so, the argument here rests

---

4 Ward (1996), *God, Chance and Necessity*, pp. 141–5, and generally pp. 61–146.

5 Ruse (1995), *Evolutionary Naturalism*, pp. 178, 185; Ruse (2003), *Darwin and Design*, pp. 309–12.

6 Ward, op. cit., *God, Chance and Necessity*, p. 143.

7 Ward, ibid, pp. 64, 145.

8 Keith Ward (2004), 'Theistic Evolution', in Dembski and Ruse, *Debating Design*, 261–74, p. 263.

9 Ward, ibid.

(despite Ward's best intentions) on too anthropocentric an account of value to be cogent. For an argument concerned with states of affairs valued by the creator, it is also too little concerned with whatever it is that confers on instrumental values the value that they have, with states of affairs, that is, whose value depends on themselves and their own nature (states of affairs that are intrinsically valuable in the usual sense), rather than on factors beyond themselves. While it is true that the world could have harboured much less value than it does, this argument (in Ward's version) could be held to add little to the Fine Tuning Argument, with its stress on the presence of what is needed if life is to be possible, and thus on what creatures have reason to want. It could also be held, unlike that argument, to focus, at least in this version, on how the world embodies the conditions necessary for human life rather than for life in general.

Ward's subsequent definition of intrinsic value (states that all sentient beings have reason to want) appears to avoid anthropocentrism and the associated problems, but at a double cost. For states that rational creatures alone have reason to want are unintentionally excluded, since other sentient creatures usually lack reasons to want them. And secondly this definition includes many states that are valuable not intrinsically but for other reasons, such as survival, strength and reproductive advantage, strangely representing them as valuable intrinsically. So it is not a suitable basis for an argument from intrinsic value, through excluding many intrinsic values and including many non-intrinsic or derivative ones.

Ward's sense of 'intrinsic value' diverges from standard philosophical usage, and seems not too far removed, where his earlier definition (what one has reason to want whatever one wants) is concerned, from what other writers call 'instrumental value', although it is closer still to what John Rawls calls 'primary goods' (things needed by anyone to carry out their purposes, whatever those purposes may be).[10] However, Ward is justified in holding that intrinsic values provide reasons for action, whether on the part of finite agents or of the creator; for if something has intrinsic value, then necessarily there are reasons for any rational agent to foster, seek or preserve it.[11] So it is worth investigating whether the concept of intrinsic value in the usual sense supports a more promising argument from value, particularly when intrinsic value is not restricted to human beings and their lives, as in Ward's earlier definition, and perhaps when it is not even restricted to sentient creatures either, as his subsequent definition assumes. Such a reconstructed argument might still support Ward's progressivism about evolution, and a version of his related claims about the creator's purposes and provision for libertarian freedom. So if the argument can be successfully reconstructed, the effort at reconstruction would be well worth while.

[10] John Rawls (1972), *A Theory of Justice*, pp. 62, 90–95.

[11] Ward, *God, Chance and Necessity*, p. 143.

Section 2: Intrinsic Value

'Intrinsic value' is standardly used of what is valuable because of its own nature, independently of other considerations. Something will be intrinsically valuable if there is reason to desire or promote or cherish it for no reason beyond itself. Intrinsic value contrasts with extrinsic value, which attaches to whatever there are extrinsic reasons for desiring or cherishing. The varieties of extrinsic value include instrumental value, attaching to artefacts devised as tools and to resources available to satisfy purposes independent of those resources; also inherent value, attaching to objects which are appreciated through a subject's observation or association with them; and contributory value, as when your perceptiveness or insight enhances our friendship. But not all value could be extrinsic; all extrinsic value depends on there being (whether immediately or at several removes) something valuable independently of it, and thus valuable of itself and because of its own nature (something such as wellbeing or flourishing), which confers value on whatever is valuable extrinsically or derivatively. Certain states of affairs, such as a person's education, can be valuable both extrinsically (because it may lead to employment) and intrinsically (because of that person's development as a person, and thus for what it is in itself). But the co-presence of extrinsic and intrinsic value does nothing to annul the distinction between the two. It should be added that, while not everyone who uses the discourse of intrinsic value interprets it in an objectivist manner, an objectivist interpretation is the one adopted here; for any other interpretation would make the reasons supplied by intrinsic value not to be universal, inter-subjective reasons, and thus, arguably, not to be reasons at all.

This kind of talk of intrinsic value should be carefully distinguished from another, to avoid possible misunderstandings. Many people talk or write as if whatever has intrinsic value has thereby certain rights such as the right to life. But this suggestion effectively invokes a different concept altogether, and no such implication of intrinsic-value discourse is adopted here. Indeed if it is states of affairs such as wellbeing that have intrinsic value, then this suggestion hardly makes sense. Besides, if there are degrees of intrinsic value (as is plausible), then it may be necessary sometimes to sacrifice one creature, despite the value of its wellbeing, to preserve the greater value of that of others, contrary to any claims about the implicit right to life of the first. Certainly rules about rights and about obligations can be reflectively grounded in the intrinsic value of whatever is at stake. But attempts to move immediately from intrinsic value to rights, even where this is apparently possible, could generate a whole ocean of potentially clashing rights, and thus whole seas of irresolvable contradictions. This may be why most writers about intrinsic value, including Ward, eschew usage of the kind mentioned in the current paragraph.

However, many people are sceptical either about the concept of intrinsic value introduced here, or about its objective character, or about our ability to apply it to the world. I have discussed these issues in *Value, Obligation and Meta-Ethics*, particularly in the chapter 'What is Intrinsic Value?', and have considered applications of this concept in the three chapters following.[12] I have also contested a range of post-modernist, subjectivist and perspectivist understandings of intrinsic value in an essay entitled 'Postmodernism, Value and Objectivity'.[13]

Grounds for scepticism can be metaphysical, epistemological or motivational, but all these grounds can be countered, albeit in ways that can only be summarized here. Thus the principal type of metaphysical objection is a version of John Mackie's argument from queerness. It protests at the idea of there being objective value-properties which, if they were to exist, would be different from everything else in the universe (or so Mackie suggests), not least in having the quality of 'to-be-pursuedness' built into them.[14] Yet there is nothing more peculiar about there being reasons for action than about there being reasons for belief, to which Mackie had no objection, and equally there is nothing odder about the desirability or value of states of affairs or of actions than there is about the credibility of beliefs or of statements. Further, if there are reasons for action, then some must be non-derivative, and something or other must enshrine them, and therewith the quality of 'to-be-pursuedness' at the same time. The suggestion that facts cannot supply reasons for action is a positivist myth, with roots such as the verificationism discussed in Chapter One above, or meta-ethical variants of the anti-realism discussed in Chapter Two. Another metaphysical objection concerns the groundedness of talk of intrinsic value in the distinction between subjects and objects, a distinction sometimes regarded as objectionable.[15] Insofar as the agents for whom or for which intrinsic value supplies reasons will also be subjects, while intrinsic value itself attaches to objective states of affairs, this distinction probably does underlie intrinsic value discourse. Yet this is an entirely unobjectionable distinction, since we all rely on it continually. Otherwise you, the reader of this book, could not distinguish between objects such as the book and subjects such as its author and such as yourself. This is not a distinctively modernist distinction which we have outgrown, as is sometimes suggested, but one embedded in the human thought of every age and period.[16]

Epistemological objections allege that there is no method (other than arbitrary intuition) by which intrinsic value can be recognized or known. This is ultimately an objection to belief in the objectivity of claims about intrinsic value. Thus Mackie suggests that all such claims rely on intuition and nothing more.[17] Yet reasoning about desirability (and thus about value) is widespread

---

[12] Robin Attfield, 'What is Intrinsic Value?', in Robin Attfield (1995), *Value, Obligation and Meta-Ethics*, 29–43; see also Chapters 4–6.

[13] Robin Attfield (2001), 'Postmodernism, Value and Objectivity', *Environmental Values*, **10**.2, 145–62.

[14] J.L. Mackie (1977), *Ethics: Inventing Right and Wrong*, pp. 38–40.

[15] Bryan Norton (1992), 'Epistemology and Intrinsic Values', *The Monist*, **75**, 208–26.

[16] See further Attfield, *Value, Obligation and Meta-Ethics*, pp. 30–32.

[17] Mackie, op. cit., p. 38.

(and indispensable), and sometimes involves moves from agreed moral stances (for example the wrongness of most killing of human beings) to shared presuppositions about value (for example the value of human life, at least where some positive quality of life is present or in prospect). That there are objective agreed moral stances is recognized not only by moral philosophers such as Philippa Foot[18] but also by such an evolutionary naturalist as Michael Ruse.[19] Yet even those who do not recognize such agreement in moral judgements are open to the force of thought-experiments designed to elicit shared non-moral values, such as Derek Parfit's Young Girl's Child example,[20] or Richard Routley's Last Man case,[21] or the case of Robert Nozick's Experience Machine, in which the unwillingness of most people to be connected indefinitely to an experience-simulating pleasure-machine serves to indicate that something other than pleasure has intrinsic value.[22] Thought-experiments such as these allow our beliefs about intrinsic value (and sometimes our related moral beliefs) to be elicited and (on occasion) rationally revised. Nor are these the only means of reasoning about intrinsic value; for example, arguments by analogy from agreed premises may establish that since the wellbeing of current people matters, so must that of future people. So epistemological scepticism about intrinsic value should be disregarded.[23]

Objections concerning motivation involve scepticism about the action-guiding role ascribed to language about intrinsic value by those who take it seriously. The mere recognition of the presence of intrinsic value as a fact, it is suggested, could not motivate or stir agents to action.[24] However, the recognition that something (such as enjoyment) is intrinsically valuable implies that there are reasons for seeking it, and thus for action, and when anyone recognizes that there are reasons for action they are liable to be influenced towards action accordingly. Thus the language of intrinsic value is capable of guiding action, not in the sense that everyone aware of such value has no option but to take corresponding action, a claim that is obviously untrue, but insofar as reasons for action have become apparent to them, and guide them accordingly, albeit reasons capable of being countered by other considerations.[25] Thus far, Ward would probably endorse these

---

18 Philippa Foot (1978), *Virtues and Vices*.

19 Ruse, *Darwin and Design*, pp. 237–9.

20 Derek Parfit (1984), *Reasons and Persons*, pp. 357–61. It was wrong of the girl not to wait until the child she bore (albeit a different child) would have been immune from an inherited disease. So the good of possible people matters of itself.

21 Richard Routley (1973), 'Is There a Need for a New, an Environmental Ethic?', *Proceedings of the XVth World Congress of Philosophy*, 205–10. The Last Man cuts down a tree. If this is wrong, as most people believe, then since no one is either harmed or benefited, nor any other creature, the tree must have been valuable in itself.

22 Robert Nozick (1974), *Anarchy, State and Utopia*, pp. 42–5.

23 See further Attfield, *Value, Obligation and Meta-Ethics*, pp. 32–34.

24 Thomas E. Hill, Jr (1984), Review of Elliot and Gare (eds), *Environmental Philosophy*, in *Environmental Ethics*, **6**, 367–71.

25 See further Attfield, *Value, Obligation and Meta-Ethics*, pp. 34–5.

defences of the concept of intrinsic value, and of an objectivist interpretation of related discourse.

It is in any case important here to emphasize that value does not consist simply in being valued, but rather belongs to what there is reason to value (or what is valu*able*). This is the central reason why subjectivist and perspectivist understandings of value are defective; for the valuable need not be valued, and what is actually valued may merely overlap with what is truly and intersubjectively valuable. (This all strengthens the case for the view that claims about intrinsic value are claims about what has such value objectively.) Relatedly (and this is where my account of the conceptual constraints surrounding intrinsic value begins to diverge from Ward's), what is valuable need not be an object of consciousness; for consciousnesses (or at least finite ones) could fail to remark or discover some of the states or entities that there is reason to value. Thus the conceptual links between value and consciousness are often exaggerated (as they may well be, for example, when Ward makes consciousness and its varieties a precondition of intrinsic value[26]). We should certainly not make the assumptions either that all forms of consciousness are valuable as such, or that things lacking consciousness can never carry intrinsic value. Health, for example, is to be regarded as desirable not only on a derivative basis, but intrinsically. But this value does not depend on awareness of its health on the part of the relevant creature, and is present whether the creature is conscious or unconscious.

This already suggests that the possibility that the health, and generally the wellbeing, of non-conscious living creatures may have intrinsic value, as well as the health and wellbeing of conscious ones. Thought-experiments such as Routley's (just mentioned) about the wrongness of the last man's action in felling a tree support this view. So does the argument of Kenneth Goodpaster that all potential beneficiaries of beneficence have moral standing, that potential beneficiaries are creatures with a good of their own, and that our value-theory should be adjusted accordingly.[27] So too does Donald Scherer's thought-experiment about the planets Flora and Lifeless; if asked whether there are reasons for preserving (rather than destroying) a planet full of organisms that grow, reproduce, photosynthesize and maintain themselves, despite lacking sentience and mobility, most people would say that there are, whereas the same people find no such reasons for preserving the aptly-named planet Lifeless.[28]

Human wellbeing (including human pleasure) is usually regarded as a central example of something having intrinsic value, but the flourishing or wellbeing of non-human creatures may be held to form another example, or rather, billions of examples. (Ward appears to endorse this view, at least with regard to the wellbeing

---

[26] Ward, *God, Chance and Necessity*, p. 143.

[27] Kenneth E. Goodpaster (1978), 'On Being Morally Considerable', *Journal of Philosophy*, **75**, 308–25.

[28] Donald Scherer (1983), 'Anthropocentrism, Atomism and Environmental Ethics', in Scherer and Attig (eds), *Ethics and the Environment*, 73–81.

of sentient creatures, but not, of course, to that of non-sentient ones.[29]) Pleasure, which is almost universally regarded as having intrinsic value, is necessarily related to attaining what one wants or values, and seems to fit Ward's definitions of 'intrinsic value' as well as more standard usage. But it is implausible (as we can learn from Nozick's Experience Machine example, and as Ward would agree) that pleasure is the only good, or (come to that) pain the only evil. Wellbeing is much broader than such hedonism would allow, and involves health and perhaps (as I have also argued elsewhere,[30] but lack the space to argue properly here) the development of the capacities essential to one's species, such as water-retention in cactuses, speed in gazelles, and practical reason in human beings. For flourishing and wellbeing are species-relative; there is no such thing as flourishing as such, as opposed to flourishing as a pig, a porcupine or as a prickly pear. They involve developing the capacities that makes one's species what it is, and in the absence of which it would be unrecognizable as such. These capacities will, of course, usually be adaptations beneficial to the species; if, however, their development has intrinsic value, then there are also reasons for agents to promote or preserve such development, even when these agents and their own species do not specifically benefit. (Even if I am wrong and flourishing should not be understood as the development or exercise of essential capacities, there would still have to be species-relative capacities in whose development or exercise it consists; and the intrinsic value of flourishing would in any case remain unaffected.)

This account of the concept of intrinsic value and of its application contrasts with Ward's, for whom, for example, power is an intrinsic value,[31] but the non-sentient realm lacks such value altogether.[32] In neither regard does Ward's understanding of the scope and limits of intrinsic value help his argument. However, the changes to Ward's account proposed here make it possible to revisit the arguments he presents from intrinsic value to God and to progressivism and purpose in evolution.

Section 3: The Argument from Intrinsic Value Revisited

The reconstructed argument from intrinsic value to God runs along lines closely parallel to Ward's argument. The world of nature is strikingly full of intrinsically valuable states of affairs in the sense just explained, such as the flourishing of creatures, and could easily have been otherwise. A rational interpretation of this phenomenon consists in the universe being the creation of a creator desirous of such value. Design on the part of a purposive and loving creator

---

29 Ward, ibid., p. 143.

30 See Attfield, *Value, Obligation and Meta-Ethics*, Chapter 4, 'Essential Capacities', 45–62.

31 Ward, *Rational Theology and the Creativity of God*, p.104.

32 Ward, *God, Chance and Necessity*, p. 143.

is a reasonable interpretation of the immense amount of intrinsic value in the world as we find it.

Intrinsically valuable states do not consist in phenomena such as power, and do not invariably comprise states of consciousness or of complexity, although consciousness facilitates states of high value such as pleasure and the development of rational, emotional and practical capacities, and many of these involve considerable complexity. They include the development and exercise of those capacities in the absence of which humans would not be recognizable as humans, but they also include the development and exercise of the essential capacities (in this sense) of other creatures. Such states are thus widespread across the realm of life (the biosphere). While the Fine Tuning Argument (see Chapter Five, Section 5 above) concludes that the universe is probably the creation of an agent that loves life, this argument concludes that it is probably the creation of an agent that loves a world of value. As Ward says of his argument, this is not an exercise in deductive or inductive reasoning, but an exercise in reasonable interpretation; it is also an appeal to the best explanation. The reasoning does not preclude the possibility that all this value is the outcome of random coincidence, but contends that a purposive explanation explains it better.

Certainly this concept of intrinsic value, besides making the flourishing of living creatures intrinsically valuable, also makes their suffering, injury, disease and premature death intrinsic disvalues. But, as has been argued in Chapter Seven above, intrinsic evils such as suffering and death (or in some cases their possibility) are instrumentally necessary (as integral aspects of indispensable systems) if intrinsic value is to be significantly attained and instantiated. Thus in a world lacking suffering, some intrinsic value could be had, for the flourishing of plants would be possible, but the kinds of flourishing that are widely considered more valuable, such as that of animals, could not. In the absence of death, it is unlikely that the world could even sustain the flourishing of plants. The concepts of intrinsic value and disvalue under consideration here thus allow evils to be taken seriously across the entire realm of life (unlike anthropocentric value-concepts), but allow of the conclusion that the whole system of nature has such high value overall as to call for a purposive explanation.

Parallel arguments can be found in the work of Holmes Rolston,[33] and, arguably, of Michael Ruse. In *Environmental Ethics*, Rolston challenges the view of Gould that the products of evolution are 'chance riches', and the view of Jacques Monod that both diversity and evolutionary development result from nothing but randomness. Randomness is not to be denied (any more than Darwinism), but 'guarantees the trial-and-error exploration of the potentialities of the system', thus generating value.[34] As Ward has more recently argued, the mutational system including the laws of nature that support it can readily be

---

[33] This has recently been argued in Mark Wynn (1999), 'Natural Theology in an Ecological Mode', *Faith and Philosophy*, **16**.1, 27–42.

[34] Rolston (1988), *Environmental Ethics*, p. 207.

seen as directional, given its products.[35] Explanations of the emergence of value in terms of causal factors alone are insufficient, adds Rolston, leaving value as an 'epigenetic anomaly'.[36] Rolston takes the argument further in *Genes, Genesis and God*. Value is not put into the world by human valuations, but is found there in biodiversity to an extent described by the sociobiologist and conservationist Edward Wilson (an atheist) as 'miraculous'.[37] Not even Darwinism, even when construed by Dennett as a 'universal acid' or reductionist methodology capable of dissolving all traditional beliefs,[38] can dissolve the amazing 'generativity' of nature. The appearance of all this biotic information, as if out of nothing, and the sheer givenness of all this value, call for a transcendent explanation.[39]

Rolston's argument is not intended to be conclusive (avoiding use of the term 'design'[40]), and his appeal to the value of biodiversity could be questioned. Given the account of intrinsic value employed here, there is no intrinsic value in diversity as such, whether biological or cultural. Yet the system that embodies biological diversity also generates a multitude of states of intrinsic value, as Rolston would agree, and partly because of this diversity at that; and, since they could not arise without this system, he ascribes to it a distinctive kind of value, 'systemic value'.[41] Whether systemic value is simply a kind of instrumental value, or whether such a classification of it is disproportionate or even meaningless (rather like the claim that the universe is instrumental to the interests of life on Earth), his stance does allow him to argue from the 'generativity' of the system (as generative of intrinsic value). This part of his argument, then, passes scrutiny, as also (as I have argued above) does his claim that value (not least natural value) is objective and not a mere function of human valuation, and thus that, in a sense, it is a given. It is open to objectors to remark that this does not establish that there is a giver. Here Rolston would agree, but would add that the phenomena of value make this interpretation more probable and more cogent than alternatives.

Michael Ruse, for his part, has an argument in favour of a 'theology of nature', which he contrasts with 'natural theology' (an enterprise that he regards as outmoded).[42] He calls this an 'argument to adaptive complexity', which concerns 'the complex, adaptive glory of the living world', something that evokes a religious response. The argument is first ascribed to Ronald A. Fisher, an adaptationist and Christian who regarded adaptation as part of God's overall plan, and to

---

35 Ward, 'Theistic Evolution', pp. 268–9.

36 Rolston, op. cit., p. 209.

37 Edward Wilson (1992), *The Diversity of Life*, p. 345, cited by Rolston (1999), *Genes, Genesis and God*, p. 362.

38 Daniel Dennett (1995), *Darwin's Dangerous Idea*, pp. 61–4, 81, 521.

39 Rolston, *Genes, Genesis and God*, pp. 362–5.

40 Rolston, ibid., p. 369.

41 Rolston, *Environmental Ethics*, pp. 186–8.

42 Ruse, *Darwin and Design*, pp. 333–6.

biologists influenced by Fisher such as E.B. Ford.[43] This argument also relates to the enjoyment to be had from studying nature, and the beauty to be found there, a humbling beauty, as attested by Canon Charles Raven, who interprets this experience as 'akin to the authentic encounter with God'. Indeed the quotation from Raven in which this is asserted forms the close of Ruse's text,[44] leaving his precise conclusion unexpressed; nor does his passage about Fisher make good the deficiency, apart from conveying Fisher's view of the 'high perfection of existing adaptation',[45] which Fisher probably regarded as exquisite enough to fulfil the divine purposes. On one possible interpretation, this 'argument' is not intended (at least by Ruse) as an argument to a creator, but as an interpretation of the natural world, with its magnificent adaptive complexity, as the abode of an immanent divinity or compelling religious or numinous force. On another interpretation, Ruse's conclusion supplies a theology of nature because the working biologist's encounter with nature's glory (whether from a Darwinist perspective or not) expresses or reveals a deity who generates all this adaptive complexity. (This interpretation coheres with many of Ruse's themes and arguments, but clashes with his reservations about religion.)

Since Ruse's conclusion is not made explicit, the argument may in any case be appropriated as parallel to the one reconstructed here. While adaptive complexity is not of course the sole location of intrinsic value, many of the kinds of biological flourishing embodying complex adaptations have such value, and this value, together with their intricacy and beauty (sources of inherent value, to use the terminology introduced above), could be held to comprise its glory. This is a glory that nature could easily have lacked, and encountering it prompts awareness of some kind of creative purpose pervading the natural world. Once again, such an awareness is not mandatory, but at least comprises a reasonable interpretation, for such an underlying purpose both illuminates all this value and glory, and explains it by making it less improbable or less incredible. (The openness of Ruse's text to this interpretation stands in interesting contrast to his rejection of the kind of progressivism shortly to be discussed, and again to his own variety of anthropocentrism, according to which creatures without consciousness lack both interests and any value not awarded by humanity.[46])

Indeed, let us for a moment cease considering the argument from value in isolation, and relate it instead to the conclusions reached above about the Cosmological and Teleological Arguments. On this basis, there are good grounds for holding that, given the phenomena of intrinsic value of the actual world, the

---

43 Ruse, ibid., pp. 160–62.

44 Charles Raven (1953), *Natural Religion and Christian Theology*, pp. 112–13; cited at Ruse, op. cit., p. 336.

45 Ruse, ibid., p. 161.

46 Ruse, ibid., p. 288. Ruse's anthopocentrism is cogently identified and criticized by Ronnie Zoe Hawkins in Hawkins (2005), Review of *Darwin and Design* in *Environmental Ethics*, **27**.2, 209–12, at p. 211.

creator, belief in whom is upheld by those arguments, could well seek such value, and supply the conditions and processes (natural selection included) necessary for its generation, together with the kind of indeterminacy necessary for the freedom of finite creatures.

Section 4: The Argument from Value to Progressivism Reconsidered

Ward's argument from intrinsic value to the purposiveness of evolution begins with a premise about the emergence, through evolution, of intrinsically valuable states including consciousness, creativity and collaboration among free creatures.[47] Of these, creativity and collaboration remain intrinsically valuable (involving, as they do, the exercise of essential human capacities). Consciousness, however, lacks intrinsic value as such, for otherwise we would be obliged to regard states of tedium and of anguish as valuable in themselves; what is valuable is not mere undifferentiated consciousness, but certain conscious activities. For many human activities anchored in consciousness are widely agreed to have such value, including the development of friendships, musical appreciation, exercises of practical reason, and emotional fulfilments; these are the kinds of activities widely held to comprise not merely living but 'quality of life'. There is also much intrinsic value outside human life, to be found, I suggest, in the development or exercise of essential capacities of the various species, some of it involving finely honed skills embodying the adaptedness of predators and of prey species alike, and some of it simply embodying the healthy development of living natural kinds. The world did not have to wait for the emergence of consciousness for states of intrinsic value to be attained. A reconstructed argument to the purposiveness of evolution can open, therefore, with all the varieties of intrinsic value manifest in the actual world, and with their contingency, in the sense that there might easily have been fewer or none at all.

The argument now proceeds along lines parallel to Ward's. Since these are just the states that a loving creator, desirous of what is intrinsically valuable or desirable, could be expected to seek to generate, there is good reason to regard evolution as having such desirable, complex states as an end or goal. The evolution of the mind, for example, will not be a mere by-product of mechanisms of survival, but integral to this purpose. Meanwhile, in the case of non-human flourishing, the argument now fits the many millennia of pre-human life much better than if human consciousness were the be-all and end-all. At the same time, there remains good reason to regard evolution as being chosen in part to supply that mix of regularity and freedom for finite creatures that makes possible the emergence of intrinsically valuable end-states based on consciousness, reflection and deliberation (such as creativity and collaboration). Thus evolution really is to be regarded as progressive and purposive. As with Ward's argument, this argument

---

47 Ward, *God, Chance and Necessity*, pp. 141–5.

is not claimed to be conclusive, but rather to supply a reasonable interpretation of the phenomena.

Such progressivism is firmly rejected by many Darwinists, maintaining that what survives in the course of time in any sphere subject to natural selection need not involve improvement. Here we should distinguish belief in progress as it has been applied to three separate spheres, human history, science and evolution.

There used to be a widespread belief that human history inevitably involves improvement (whether gradual or revolutionary), and that history complies with laws of progress. For some this belief had a religious basis and a purposive content; other adherents were committed to a secular metaphysic, for which both intellectual and social progress was guaranteed. This belief originated with Enlightenment figures such as Condorcet, and was held in a variety of forms by Kant and Hegel, by Saint-Simon and Comte, and by Marx, Darwin and Spencer,[48] but was increasingly discarded during the twentieth century. The impossibility of finding laws with which human behaviour might comply has been one of the grounds for the demise of this belief; world wars and massacres have been another. (The unavailability of laws of history further suggests that human history is not in any case a sphere to be understood as subject to natural selection.) Meanwhile, versions of theistic belief that uphold some form of progressive view of history (for which, despite human lapses, 'God is working his purpose out as year succeeds to year') are dependent on revelation, and thus fall outside the scope of the present work. In any case, the defence of progressivism about evolution does not depend on progressivism about history; for the former does not imply the latter. It would not be difficult to imagine self-determining creatures such as human beings thwarting, at least temporarily or in part, the fulfilment of any purposes that may characterize the pre-human and non-human realms, even though such moral evil was argued above (in Chapter Seven above) to be compatible with God's goodness, and worlds in which its occurrence is a possibility to have greater value as such than those in which it is impossible. Accordingly, no stance needs to be taken here on historical progress, beyond what is said below about progress in the sciences and the humanities.

Progress in science can be considered differently, for it is widely held that the work of scientists such as Darwin and Mendel comprised advances on previous understanding. But evolutionary naturalists such as Ruse, who firmly maintain that there is no progress in evolution, are ambivalent on whether there is progress in (what they regard as) the cultural evolution of the scientific community.[49] Some such as David Hull adopt an evolutionary view of science itself, and maintain that success in science consists simply in getting one's ideas accepted by others in the scientific community.[50] For his part, Ruse, while endorsing a social-practice view

---

48 See John Passmore (1970), *The Perfectibility of Man*; Robert Nisbet (1980), *History of the Idea of Progress*.

49 Ruse, *Evolutionary Naturalism*, pp. 175–85.

50 David Hull (1988), *Science as a Process*.

of science, and accepting that progress is not inevitable, finds Hull's relativizing view hard to reconcile with the actual progress that is sometimes apparent.[51] Meanwhile critical realists stress that science at least aspires to truth, that it sometimes has good grounds for claiming the truth or approximate truth of its theories, and that the criteria of truth and of survival-value may well diverge; there is no reason to credit a theory just because it has survived.[52] Here, Plantinga's criticisms of Darwinian accounts of knowledge and truth,[53] discussed in Chapter Six, Section 3 above, become relevant again, or would do if science is assumed to be a realm governed largely by natural selection as opposed to rationality.

But the problem for Hull and Ruse may well be this very assumption. This is what inclines them to hold that because there is no progress in evolution, there can be none in the evolution of cultural practices such as science either. (They would of course have to say the same about the social sciences and the humanities as well.) Once this assumption is discarded, we become free to recognize progress in science (and even in the humanities), without prejudice to issues of progressivism about evolution. This recognition is of course open to atheists and agnostics as well as believers; for there could be scientific progress without any transcendent being intending or generating it. It is also open to theorists who continue to interpret the sciences and the humanities as socially embedded practices, since social practices need not be construed as invariably the site of a competition or power struggle between rival ideas (sometimes interpreted as memes), regardless of their merits. Indeed cultural practices require different methods of interpretation from the subject-matter of natural science.[54]

We now return to the issue of progress in evolution. Here, the reasons for detecting cultural progress, as in the sciences and the humanities, are absent. Indeed, as far as the theory of natural selection is concerned, there is no reason to regard what survives as better, except in the sense of being adapted to survive in a particular immediate environment. Equally, there is no reason to regard the evolution of the human species as progress simply in adaptationist, evolutionary terms,[55] although if we accept that humanity can itself make progress in the arts and the sciences, we need to qualify the way in which we express this view. However, the stance that Ward characterizes as 'theistic evolution' is not an impossible or contradictory one, and by taking into account intrinsic value supplies a criterion of progress, and at the same time a ground for detecting purposiveness in evolution.

---

51 Ruse, *Evolutionary Naturalism*, pp. 178–9.

52 Roger Trigg (2002), *Philosophy Matters*, pp. 80–81; Michael C. Banner (1990), *The Justification of Science and the Rationality of Religious Belief*, pp. 35–7.

53 Alvin Plantinga (1997), 'Methodological Naturalism', *Perspectives on Science and Christian Faith*, **49**.3, 143–54.

54 This is further explained in R. Hooykaas (1974), 'Nature and History', in Open University Course Team for Science and Belief, *The New Outlook for Science*, pp. 20–23.

55 Ruse, *Evolutionary Naturalism*, p. 185.

Whether or not Ward's argument (or the reconstructed version advanced here) is accepted as an independent argument, it at least becomes hard to resist if the other theistic arguments defended above are accepted. Further, if Ward's argument from intrinsic value to God is accepted, then the argument to purposiveness in evolution becomes readily acceptable as well; for a creator who institutes natural selection and who loves intrinsic value including that present in the exercise of deliberation and discretion among free creatures is unlikely not to have the various kinds of intrinsic value, including this one, among his or her purposes. Besides, belief in natural selection does not of itself preclude belief in evolutionary progress or purposiveness; for, as was remarked in the previous section, even random mutations could form part of a directional system.[56] What precludes this belief are various metaphysical interpretations of Darwinism such as materialism (whether of the moderate, 'blank-paper' or of the 'gene-machine' variety) and the kind of metaphysical naturalism discussed in Chapter Six above. 'Mind-first' Darwinism (the kind of stance actually held by Darwin for much of his life) is compatible with such belief, as long as teleological explanations of particular creatures or phenomena are not smuggled back in through the back door, and as long as the general operation of natural laws (including natural selection) is accepted.

The problems for Ward's theistic evolution stance lie rather with the apparent suggestion that the purpose of evolution lies either with consciousness or with complexity,[57] as if there were no intrinsic value in the non-conscious realm or in the flourishing of relatively simple creatures such as unicellular organisms, or others that have changed little since life first emerged. Such a stance is not far removed from the metaphysical anthropocentrism that makes the emergence of either humanity or at least sentient creatures the sole purpose of creation, ignoring the intrinsic value present in all creatures whether sentient or insentient, complex or relatively simple. But there is no need for a theistic evolutionist to say this, or to ignore such value. Relatedly, there is no need for those who reject Ward's distinctive (sometimes sentientist, sometimes anthropocentrist) version of theistic evolutionism also to reject the greater intrinsic value present in complex capacities, such as the development of conscious relationships;[58] indeed the

---

[56] Ward, 'Theistic Evolution', pp. 268–9; see also D.J. Bartholomew (1984), *God of Chance*.

[57] Ruth Page (2004), 'Panentheism and Pansyntheism: God in Relation', in Clayton and Peacocke, *In Whom We Live*, 222–32. A parallel objection, from a very different stable, to making humanity the unique goal and purpose of evolution can be found at Dennett, *Darwin's Dangerous Idea*, p. 56, and in different dress at Dawkins, *The Blind Watchmaker*, pp. 262–3.

[58] I have argued for a theory of degrees of value in Attfield, *Value, Obligation and Meta-Ethics*, pp. 79–94. Whether or not that argument succeeds, it is widely agreed that the development of some capacities, such as the capacity to enter into conscious relationships, is of greater intrinsic value than that of others, such as respiration, indispensable as the latter is for the former.

boundaries of complex capacities do not in any case correspond to the boundaries between species.[59]

Thus the purpose underlying evolution could be the generation of intrinsic value, as the reconstructed argument claims, without any suggestion that this purpose was unfulfilled until either sentient creatures or human beings first appeared. Evolution, as Midgley has argued, need not be regarded as a Lamarckian or Spencerian escalator leading up ever higher to humanity, and can instead be regarded (as by Darwin and Midgley) as a bush, radiating forms in multiple directions,[60] each (we can add) capable of bearing intrinsic value. (There could thus be progress and directionality not just in one but in many directions.) It would at the same time be possible to recognize, as I have argued elsewhere (but lack space to argue here), that there is greater intrinsic value in the exercise of more complex capacities (such as the development of conscious relationships) than in that of relatively simple ones,[61] and thus that such development among conscious creatures could fulfil the underlying purpose more fully (without, as Ward suggests, fulfilling it uniquely). This understanding of intrinsic value recognizes degrees of value and the role of consciousness and of complexity, without an anthropocentric dismissal of the intrinsic value in the lives of creatures that are neither complex not conscious, and thus of much of the glory of the radiating bush of the biosphere.

Section 5: Purpose, Immanence and Panentheism

The argument of the preceding sections about a purpose underlying natural processes such as natural selection suggests that these natural processes are the way in which God creates the various life-forms. Belief in creation certainly means that creatures are dependent on God at all times, but it also means that God bestows on them their form; and this is done not in a single instant, but step by step in the course of evolution. While this is a timeless bestowing (if, that is, the argument of Chapter Four above is accepted), it is achieved through created temporal processes continually generating new forms and species. These processes include physical and chemical ones, through which the elements are made from subatomic particles, and galaxies, stars, comets and planets from the dust and debris of an expanding universe. They also include geological ones, through which continents, mountain chains and oceans emerge from the forces of plate tectonics, and supply the environments for biological processes such as

59 Robin Attfield (1994), *Environmental Philosophy*, pp. 203–20.

60 Mary Midgley (1985), *Evolution as a Religion*, pp. 6, 69–70.

61 Attfield, *Value, Obligation and Meta-Ethics*, pp. 79–94.

natural selection on life-bearing planets such as Earth. Thus God creates through naturalistic processes, in which 'things make themselves'.[62]

If creation operates in part through natural processes, God is to be seen not only as transcending the natural order but also as immanent in it. While this immanence will not be confined to biological processes such as natural selection (and may extend to the processes of human culture and history), the unseen operation of creative purposes across evolutionary history will play a significant part in its overall drama. God will be present in the evolving world rather as a composer is present as his or her intentions are expressed during a performance of a musical work such as a symphony.[63]

Such a theological interpretation of evolution was given early expression by Aubrey Moore not long after Darwin's death. By comparison with a theory of special creation, the evolutionary theory is 'infinitely more Christian' (or, as Moore could have said, more consistent with theism in general). 'For it implies the immanence of God in nature, and the omnipresence of his creative power'; and its opponents 'seem to have failed to notice that a theory of occasional intervention implies as its correlative a theory of ordinary absence'.[64] Moore developed his advocacy of divine immanence and opposition to a deism of occasional interventions in his contribution to *Lux Mundi* (1891).

> The one absolutely impossible conception of God, in the present day, is that which represents him as an occasional visitor. Science has pushed the deist's God further and further away, and at the moment when it seemed as if He would be thrust out altogether, Darwinism appeared, and, under the disguise of a foe, did the work of a friend … Either God is everywhere present in nature, or He is nowhere.[65]

Not every form of deism would need to be discarded if Moore is right. Thus a deism grounded in natural theology to the exclusion of revelation, such as that of the English deists like John Toland and Matthew Tindal, and of French deists such as Voltaire and Rousseau, would not be affected, particularly in versions that related creation not just to cosmic origins but also to the whole trajectory of

---

62 Peacocke (2004), 'Articulating God's Presence in and to the World Unveiled by the Sciences', in Clayton and Peacocke, 137–54, p. 142. The idea of God making 'things make themselves' comes from A.M. Farrer (1964), *Saving Belief*, p. 51; the phrase originated, appropriately, with the Christian evolutionist Charles Kingsley, in Kingsley [1863] (1930), *The Water-Babies*, p. 248. It reappears in the process that Steven Rose calls 'autopoiesis': see Rose (1997), *Lifelines: Biology, Freedom, Determinism*, p. 18.

63 Arthur Peacocke (1985), 'Biological Evolution and Christian Theology – Yesterday and Today', in John Durant (ed.), *Darwinism and Divinity*, 101–30, p. 124.

64 A.L. Moore (1889), *Science and Faith*, p. 184.

65 A.L. Moore (1891), 'The Christian Doctrine of God', in Charles Gore (ed.), *Lux Mundi*, 41–81, p. 73.

time.[66] But Moore has a strong point against deisms whose deity is either absent or abandons creation to its own devices or intervenes occasionally to supply any gaps left by science. Such forms of deism fall short of classical theism, with its stress on the dependence of the material or physical world at all times upon God. Once evolution is seen as an aspect of creation, and these processes are recognized to be largely coextensive, an immanentist theology becomes hard to resist. The theologian and naturalist Charles Raven (mentioned in Section 3 above) was among its twentieth-century advocates.[67]

Darwinism thus emerges as a potential ally of theism (and not merely as a compatible stance, as argued by Gould and Ruse, let alone as a universal acid, as suggested by Dennett).[68] If evolution by natural selection was chosen to fulfil the creator's purposes, then Darwinism not only coheres with theism but redirects theism towards immanentist versions that recognize God's ongoing creativity. At the same time the apparent remoteness of the creator of classical theism turns out to be an illusory construct. The reconstructed arguments from intrinsic value support just such a combination of creation and evolution, and do so on the basis of the world's intrinsic value, embodied (in part) in its adaptive complexity.

Arthur Peacocke further depicts such an immanentist theology as 'panentheism', a position defined as 'the belief that the Being of God includes and penetrates the whole universe, so that every part of it exists in Him, but (as against pantheism) that His Being is more than, and is not exhausted by, the universe'.[69] This position was briefly discussed in Chapter Four, Section 1 above. It corresponds closely to Paul's description of God as the one 'in whom we live and move and have our being'.[70] Peacocke is not supportive of the panentheism often found in process theology, for which God is subject to and dependent on creation, but rather of a form of panentheism (such as that of Jürgen Moltmann) which fully respects the distinction between God and the world, and also God's unqualified transcendence,[71] thus holding together 'both the transcendence and the immanence of God in relation to the world'.[72] As was mentioned in Chapter Four above, panentheism, through celebrating the natural world, is more prone to recognize the intrinsic value of its constituents than classical theism (for example that of Barth) has sometimes been able to do (although classical theism and the Moltmann/Peacocke form of panentheism remain compatible positions:

---

66 Robin Attfield (2004), 'Rousseau, Clarke, Butler and Critiques of Deism', *British Journal for the History of Philosophy*, **12**.3, 429–43.

67 Peacocke, 'Biological Evolution', pp. 112–13.

68 Dennett, *Darwin's Dangerous Idea*, pp. 61–4.

69 Peacocke, 'Articulating God's Presence', p. 145; F.L. Cross and E.A. Livingstone (eds) (1974), *The Oxford Dictionary of the Christian Church*, p. 1027.

70 Acts, 17: 28.

71 Peacocke, 'Articulating God's Presence', pp. 145–46; Jürgen Moltmann (1985), *God in Creation*, pp. 88–9.

72 Peacocke, *Theology for a Scientific Age*, pp. 208–9.

see below). At the same time, Peacocke seems right to resist the contradictions that would result if (as in the panentheism of much process theology) the creator were understood as dependent on or in any way subject to the creation, or if, as Craig suggests, panentheism implied the eternity or necessity of the world.[73] Nor need panentheism regard the world as somehow contained or inherent in God, as Philip Clayton's version of panentheism commends.[74]

Keith Ward is reluctant to endorse panentheism, because of (what he considers) its tendency to think of the world as God's body, as pantheists also do.[75] However, panentheists, not being pantheists, are free to deny that the world is God's body, and Ward's remarks about the fulfilling of the Creator's purpose in evolutionary history cohere well with the panentheism just presented. Thus the goal of evolution is the generation of 'beings who understand how to generate worthwhile states creatively, and who find fulfilment in contemplating them. In that respect, such beings are images of the divine being … .'[76] While this claim is expressed in an unduly anthropocentric form, and both intrinsic value and the goal of evolution are also to be found in the flourishing of other, less contemplative creatures, it could well be held to capture a significant element of the world's value, and one that panentheists can and should recognize. The same qualifications, and the same qualified endorsement, apply to his evocative claims that 'the cosmos is an image of God insofar as it comes to embody self-awareness and free agency', and that the goal of the cosmos 'is likely to lie in the realisation of a conscious relationship of the cosmos to its creator'.[77] Panentheists can add to this account the value present in flourishing non-human lives, lives with a more circumscribed freedom than the ones celebrated by Ward, and generally lacking awareness of self or creator; for, on the above reconstructed account, the purpose underlying evolution was already being realized long before the emergence of humanity. (Further, that purpose would not have to be regarded as defeated even if humanity were eventually to be superseded by other species in the course of evolution in the future.)

Owen, whose account of classical theism, Platonism, pantheism and panentheism was discussed and largely adopted above in Chapter Four, rejects panentheism as contradictory. But the panentheism that he rejects is the panentheism of process theology, as in the thought of Charles Hartshorne and Schubert Ogden; as he insists, God cannot both be changeless and changing, infinite and finite, eternal and temporal.[78] Here he seems to make his case.

---

73 Copan and Craig (2004), *Creation Out of Nothing*, pp. 13–14.

74 Philip Clayton (1997), *God and Contemporary Science*, pp. 47, 223.

75 Keith Ward (2004), 'The World as the Body of God: A Panentheist Metaphor', in Clayton and Peacocke, *In Whom We Live*, 62–72. This is, however, the stance of Clayton, *God and Contemporary Science*, pp. 262–4.

76 Ward, *God, Chance and Necessity*, p. 186.

77 Ward, ibid., p. 187.

78 Owen, *Concepts of Deity*, pp. 83, 88, and 147.

However, his earlier definition of panentheism (which is identical with that used by Peacocke and cited above) accommodates also the panentheism of Moltmann and Peacocke, which can escape any such charges of contradiction. By actualizing and employing finite and temporal processes of creation, God does not cease to be infinite or eternal (see Chapter Four, Section 4 above), much less become dependent on the created order, and by making creatures make themselves, and thus carry through the creative process, God does not cease to be changeless; nor need classical theism say otherwise. We are not tempted to say, even of a human composer such as Beethoven, that he grows or develops as a work of his is performed, even in cases where the performance occurs during his lifetime, simply on the basis that the delivery or execution of his intentions is spread across time. If there is such growth, it is not due to the fact of the performance, but to extraneous factors such as the composer's contingent reaction to it. All the less should we be inclined to say that because God employs temporal processes of creation and is thus immanent in the world, the creator grows, develops or changes. This kind of panentheism, then, is not immersed in the kind of contradictions alleged by Owen. While considerations of space prevent a fuller consideration of Peacocke's accounts of divine action and causation, there is no reason to suppose that other contradictions or problems would make such panentheism untenable. Indeed Philip Clayton has argued that panentheism overcomes apparent problems about divine agency more readily than classical theism can seem to do in its absence (particularly if adherents of classical theism deny that God can employ finite processes);[79] but these issues cannot be further investigated here.

The reconstructed arguments from the world's intrinsic value to God and to the purposiveness of the evolutionary process can thus be regarded as adding support to an immanentist theology and to a version of panentheism. Darwinism was recognized to support an immanentist theology by Moore soon after Darwin's death, and the resulting theology has more recently been developed and better articulated as panentheism by Moltmann and Peacocke. The arguments from intrinsic value introduced by Keith Ward prove relevant to this stance, but their relevance becomes much more apparent when they are reconstructed as above. Even if they cannot conclusively vindicate the case for panentheism, they go some way towards supporting belief in God as a creator desirous of value, and in evolution as a process designed to generate such value. (Indeed Peacocke first substantially presented the case for panentheism in *Creation and the World of Science*, a work which, without actually employing the argument from intrinsic value, well related ecological value to theistic insights.[80])

---

79 Philip Clayton (2000), *The Problem of God in Modern Thought*, pp. 504–5.

80 Peacocke (1979), *Creation and the World of Science*; see in particular pp. 255–318. Peacocke has gone on to remark (at 'Biological Evolution', p. 126) that God expresses meanings in his creation, and that people find that in some sequences of events God unveils his meaning more than in others; but I cannot discuss revelation, as opposed to its possibility, here.

In the current chapter I have attempted to construct a more consistent, non-anthropocentric version of Ward's arguments from intrinsic value to God and to purpose in evolution, and also to adduce them in support of Peacocke's and Moltmann's panentheist case. These reconstituted arguments from value complete Part II, supplementing the case presented in Chapters Five to Seven for theism having good grounds, and for its being consistent with Darwinism and with moral and natural evil; and thus support belief in creation, as depicted (at the beginning of Part II) in Chapter Four, as well as the high value of the created order and the creator's good purposes.

PART III
Evolution and Meaning

Chapter Nine

Meaning, Evolution and Stewardship

Section 1: Introduction

In the final part of this book, meaning, in the senses both of linguistic meaning and of meaningful action, is related to evolution; the links between both evolution and meaningful action on the one hand and stewardship on the other are investigated; and the metaphysics and ethics of stewardship are explored.

In the current chapter, discussion of a recent suggestion concerning the near equivalence of meaning and value generates a renewed investigation (supplementary to that of Chapter One above) of the concept of meaning and neighbouring concepts, including meaningful action. This is followed by a consideration of a Darwinian and deterministic account of the possible evolution of meaning and of related capacities, of which some elements are found to throw light on the nature and development of understanding and meaning, even if gene-machine versions of Darwinism and deterministic accounts of culture should be rejected. In a further section, meaningful action is related to meaningful work and meaningful lives, and problems raised concerning prospects for such lives by Darwinism are appraised, while in the final section meaningful action is shown to be epitomized by (among other roles and activities) adoption on the part of human beings of the role of steward, trustee or guardian of the natural world, and thus of the products of evolution on our planet. The presuppositions, metaphysics and ethics of stewardship are considered in the final chapter.

Section 2: Meaning, Value and Meaningful Action

Since the previous chapter focussed on intrinsic value, and meaning is the focus of this one, my starting-point is an intriguing essay that seeks to explain intrinsic value by relating it to meaning.[1] In the course of the discussion, the author, Bill Hook, considers Daniel Dennett's account of the birth and evolution of meaning, an account which is discussed below in the coming section. Hook's specific conclusions about the location and bearers of intrinsic value cohere well with the biocentric stance adopted above in Chapter Eight of this book. But his ambitious attempt to relate value to meaning in the sense of the realization of

---

[1] Bill Hook (2003), 'Intrinsic Value: Under the Scrutiny of Information and Evolutionary Theory', *Environmental Ethics*, **25**.4, 359–73.

information by an intelligent organism or 'reader' calls for scrutiny and reflection, particularly where the concept of meaning is concerned.

Does value, and does even intrinsic value, require the presence of an intelligent organism or creature in order to be realized? According to Hook, intrinsic value is carried by entities that are both realizations and realizers of meanings (and thus creatures with some measure of intelligence), whereas extrinsic value belongs to things that are realizations only (such as artefacts or works of art).[2] Hook's instincts are on the right lines here about artefacts and works of art, but this stance, on the face of it, confines intrinsic value to the realm of intelligent and thus self-moving, conscious creatures. Yet he also holds (in a context concerning intrinsic value) that 'value permeates the biosphere, from the simplest single cell organism to humans, who, through the gift of accumulated design, are uniquely situated to make it decisively meaningful'.[3] Deferring for a moment issues of meaningfulness, this position becomes coherent only if single cell organisms, and thus all organisms, are held to be readers, embodying and replicating design, and thus to some degree intelligent. This is probably what Hook intends, since Dennett's account of the accumulation of design is best interpreted in this sense.[4] But since the concept of intelligence is normally restricted to animals with some minimal degree of consciousness, and this restriction excludes the majority of living creatures, biocentrists (including Hook) would be better advised not to limit the scope of intrinsic value in this way, and explicitly to include non-conscious organisms (if not also the development of their essential capacities) among its bearers.

While the bearers of intrinsic value could still be realizations and realizers of genetic codes, any apparent conceptual links between value and meaning now begin to seem overstretched. Sometimes, admittedly, 'having meaning' simply amounts to 'significant' or 'valuable'; but 'meaning' usually has a broader sense. Thus meanings (in the sense of linguistic meanings) indifferently concern what is valueless as well as what is valuable, and cannot be mapped against values. Again, the meaningfulness of actions and events applies not only to values but also to significant evils or disvalues. Meaningful events, indeed, often have negative value, such as news of a family disaster. Thus not all meanings can be realized as values, and not all values attach to creatures capable of grasping or expressing meanings. So even where intrinsic value involves realization of possibilities, it may be better not to represent the relevant possibilities as meanings, as opposed to inherited capacities. (Dennett's account of how meanings and intelligence gradually evolve by natural selection from more rudimentary forms of design will be considered in the coming section; since it does not invoke any explicit value-theory, it need not be considered here.)

---

2 Ibid., pp. 371, 373.

3 Ibid., p. 372.

4 See Daniel Dennett (1995), *Darwin's Dangerous Idea*, pp. 68–73, 204–5, 401–3.

Yet, as Hook makes clear, meaning requires readers.[5] 'Readers' here has to be fairly generously interpreted (if not as generously as in Hook's probable application of the term to all organisms), as individuals capable of grasping meaning include intelligent hearers, whether or not literate, not excluding the very young, the semi-comatose, and some domestic animals; but the class of readers is very much a subset of intelligent animals (see the Popperian creatures of the next section). Meaning, as Dennett agrees, in any case depends on minds, and vice versa: 'Philosophers commonly agree, for good reason, that meaning and mind can never be pulled apart, that there could never be meaning where there was no mind, or mind where there was no meaning.'[6] Even if the minds of mice and guinea-pigs are not equal to linguistic meaning, they remain equal to interpreting signs of danger and adjusting their behaviour accordingly, and this capacity could reasonably be understood as a recognition or reading of meaning. Yet far from all purposive animals have sufficient powers of perceptual interpretation to be regarded as readers of meanings, even if such powers can be acquired by their evolutionary descendants. (The sea-slug *Aplysia* is cited by Dennett as an example of a creature subject to simple conditioning in the pursuit of its own good, but lacking foresight and capacities for pre-selection among options for behaviour.[7])

Besides, meaning requires communicators as well as readers, and readers incapable of communication have only a tenuous grip on the world of meaning. While communication may not be the only role of language, or the only purpose that it evolved to satisfy, any account of meaning must obviously recognize its centrality. Capacities for meaning, then, involve the capacity to generate signs and symbols as well as to interpret them. And while creatures that receive information and perceive its significance may be at the threshold of grasping meanings, participation in communication clearly involves a greater mastery of one or another semantic system, and thus semantic skills. This is a much greater accomplishment than being a mere reader of signs, whether in the sense of intelligent interpretations of perceptions, or in the more rudimentary sense (which may be Hook's sense) of interpreting one's genetic code simply by being or becoming a phenotype. (As an exercise of an essential human capacity, participation in communication may also be held, on the basis of the value-theory of Chapter Eight above, to carry intrinsic value. So it also would in the lives of cetaceans if this is an essential capacity of dolphins or of whales.)

Communication of meanings further enables some utterances to be distinctively significant or meaningful ones, resonating with the central priorities or the deeper concerns (see Chapter One, Section 1 above) of speaker or writer, intended hearer or reader, and/or chance audience or readership. Actions too can be meaningful through bearing a parallel relation to priorities or concerns

---

5 Hook, op. cit., pp. 369–70.

6 Dennett, op. cit., pp. 204–5.

7 Dennett, ibid., p. 376.

(and so indeed can lives); for the agents of meaningful actions (and lives) are able to identify with their deeds and roles, and sometimes to regard them as having significance in a larger scheme or in the shared practices of communal life.[8] While meaningfulness in this sense is obviously not confined to linguistic items or acts (indeed musical themes and works too can be meaningful, as also can historical episodes in a related sense), this kind of meaningfulness is only intelligible against a background of shared semantic capacities, capacities for meaningful communication. Accounts of meaning should accordingly provide for the dimension of meaningful communication and action. Nothing less will make sense of Hook's cogent remark that humans 'are uniquely situated to make it [almost certainly, the value that permeates the biosphere] decisively meaningful',[9] even if he may not have had in mind either meaningful actions or the human capacities for self-awareness and self-creation.

Another implication (already drawn in Chapter One above) is that our speaking both of the meaning of expressions and of the meaning of actions and of life is no coincidence, for the relevant senses of 'meaning' are interconnected; for meaningful actions are only intelligible against a background of shared semantic capacities. Yet meaningful actions and lives are found only among agents with a concept of themselves as agents; for in the absence of this, neither actions nor lives can be regarded as meaningful.[10] This conceptual requirement needs to be borne in mind when we consider how, if at all, meaning and understandings of meaning could have evolved, the theme of the coming section.

Section 3: The Evolution of Design, Intelligence and Meaning – Dennett's Tower

In this section, Dennett's account of the evolutionary development of design, intelligence and meaning is used to assist reflection on the evolution of intelligence and of the kinds of understanding related to meaning and meaningful action. This is the narrative of accumulating design across evolutionary history regarded by Hook (see the previous section) as facilitating meaning and meaningfulness in human life. No pretence is made to study here the development of the brain, either in terms of its evolution across millennia or of evolution from embryo to mature individual. I am concerned rather with the evolution of the mental capacities necessary for intelligent behaviour, linguistic communication and meaningful action. Despite large divergences from Dennett, I believe that much can be learned from his narrative of the evolution of intelligence and meaning, or of their antecedents, through natural selection.

---

8 David E. Cooper (2003), *Meaning*, pp. 30–31. Following Dilthey, Cooper calls these shared practices 'Life' (p. 30).

9 Hook, op. cit., p. 372.

10 John Cottingham (2002), *On the Meaning of Life*, pp. 21–2.

Nevertheless, Dennett's frame of reference should be borne in mind. He is seeking to defend what Richards calls a 'gene-machine' account[11] of the evolution of mind, of meaning and thus of human beings, in which naturalistic, adaptationist and deterministic processes of natural selection (what Dennett calls 'cranes') are sufficient to explain the phenomena of human life; for which human beings are survival machines generated for the sole purpose of their survival by our genes; and in which any beliefs or behaviours unexplained by genes are explicable by the impacts of a further set of non-rational, self-replicating entities, the memes. Any other kind of explanation introduces 'skyhooks', which (as was mentioned above in Chapter Six) are one and all illusory, whether they expressly invoke a supernatural designer or, as in Chomsky's and Gould's theories, non-adaptive linguistic capacities that are not explicable by Darwinian processes.[12] Meanings, by contrast, develop through an accumulation of design. This is not the kind of design generated by a designing mind, but the kind present in the increasingly complex and sophisticated products of natural selection. Rather than respond with a head-on critique of such metaphysical Darwinism (along lines, perhaps, of the critiques of Dawkins from Ward and McGrath, depicted above in Chapters Six and Seven[13]), I prefer to present central aspects of Dennett's fascinating narrative; for if human beings (and thus their linguistic practices) are among the fruits of evolution, then it is apparently possible for their language and meanings also to evolve; and Dennett offers to show how this could have taken place.

According to Dennett, there is a pyramid of design, matching the pyramid of creatures credited by William Paley, with minds and their art and artefacts at its pinnacle.[14] This pyramid arises through an accumulation of Darwinian processes, which derive from the codings present in clusters of nucleotides,[15] and thus from our selfish genes.[16] The pyramid is now presented as a tower, with multiple tiers or storeys. At the lowest level of Dennett's 'Tower of Generate-and-Test',[17] there are Darwinian creatures, generated by their genes to display a random array of configurations and behaviours, which were 'field-tested' against environments; many were eliminated but 'the best designs survived' and multiplied.[18] Most were hard-wired to behave as they do and must do, but the behaviour of some proved susceptible to conditioning and reinforcement of behaviours beneficial to the 'agent' ('Smart Moves'), and many of these were among those selected for survival.

---

[11] Janet Richards (2000), *Human Nature after Darwin*, pp. 55–6.

[12] Dennett discusses the relevant theories of Chomsky and Gould at *Darwin's Dangerous Idea*, pp. 384–93.

[13] Ward's and McGrath's critiques of Dawkins are variously presented in Chapter Six, Sections 3 and 4, Chapter Seven, Section 4, and Chapter Eight, Section 1 (above).

[14] Dennett, op. cit., p. 70.

[15] Dennett, ibid., pp. 204–5.

[16] Dennett, ibid., p. 426.

[17] Dennett, ibid., p. 373.

[18] Dennett, ibid., p. 374.

Dennett calls this subset 'Skinnerian creatures' (creatures of 'conditionable plasticity'), which form the second level of his Tower.[19] But since such creatures are liable to be killed by their early errors, Darwinian processes selected for a subset of this subset which exhibited pre-selection of possible behaviours, rejecting misguided moves beforehand, and thus acting with foresight. These creatures use information from the outer environment to equip their 'inner environment' to facilitate this pre-selection. Dennett calls this third level of his Tower 'Popperian creatures', since Popper once memorably remarked that foresight 'permits our hypotheses to die in our stead'. Popperian creatures survive because they make 'better-than-chance' early moves.[20] An example of such creatures might be herring-gulls veering in flight as they approach a cliff; but human beings are to be regarded as Popperian creatures too.

The next storey of Dennett's Tower comprises 'Gregorian creatures', named after the psychologist Richard Gregory. The inner environments of Gregorian creatures are informed by the designed portions of the outer environment (such as scissors, hammers, sticks and words). Such creatures include those chimpanzees that use tools; others, however, lack the intelligence either to recognize a tool or to use it intelligently. Dennett holds that tool-use involves (and also confers) a significant increase in intelligence, and presumably accepts that being a Gregorian creature typically confers survival advantage.[21] Tools include what Gregory calls 'mind-tools' (words), which embody the teamwork of numerous brains, and enable these creatures to think better about what they should think about next, plus numerous other kinds of internal reflection.[22] Here it may be observed that although linguistic creatures are Gregorian creatures, far from all of the latter have language. Since language facilitates a significant development of intelligence, linguistic creatures could be regarded as a further storey of the Tower (and are effectively treated so by Dennett, although he holds back from bestowing on them a distinctive name; the obvious name 'Chomskyan creatures' would perhaps go too greatly against the grain).

Linguistic creatures, adds Dennett, generally have such advantages as a 'theory of mind module' designed to generate second-order beliefs (beliefs about the beliefs and other mental states of others), a capacity lacked by some autistic people, and they also have the kind of foresight allowing of conscious reflection on the potential outcomes of possible current actions.[23] Thus having language (and the understanding of meanings that goes with it) allows us to use mind-tools 'in the structure of deliberate, foresightful generate-and-test known as science'. This comprises the final tier of the Tower of Generate-and-Test.[24] The Tower

---

[19] Dennett, ibid., p. 374.
[20] Dennett, ibid., p. 375.
[21] Dennett, ibid., pp. 377–8.
[22] Dennett, ibid., p. 378.
[23] Dennett, ibid., p. 379.
[24] Dennett, ibid., p. 380.

puts our minds on a different plane from those of our nearest relatives among the animals. Human brains are (so to speak) joined together through language into a single cognitive system that dwarfs all others, insofar as each human brain is the beneficiary of the cognitive labours of the others.[25] (But since some non-humans learn language, many others respond to it, and not all humans have these accomplishments, this paean to language-based culture risks association with the kind of essentialism of species that Dennett, as an anti-essentialist, is committed to eschewing and rejecting.)

One of Dennett's claims is that language evolved through natural selection by such Darwinian pathways, and that language and meaning are thus explicable on this basis;[26] this is discussed below. He is not, however, claiming that we occupy only the upper tier of his Tower, but rather that linguistic creatures are also, for some purposes, Darwinian, Skinnerian, Popperian and Gregorian, as well as linguistic. For some of our behaviour is not subject either to control or to reinforcement, some embodies conditioned reflexes, some unreflective foresight, and so on; we embody all the levels of understanding, intelligence and intentionality (or their absence) present in the Tower, since the upper storeys are not new beginnings but evolve from and are constructed out of earlier and lower levels. Granted that human beings are in some sense outcomes of evolution, this is a valuable insight into our mix of levels of intelligence and understanding, and helps explain the different layers of meaning that we grasp or fail to grasp. For our ancestors were learning to read the meaning of situations, adapt their behaviour and exercise rudimentary foresight long before they were human.

Section 4: Some Criticisms of Dennett's Tower

However, Dennett writes as if science were the ultimate design achievement, and the pinnacle of the Tower. But only some science uses the second-order beliefs that he mentions, and many non-scientific pursuits use them with much greater sophistication and insight. Therapeutic counselling, for example, is not science (despite sometimes being based on science), but involves a high level of understanding of others' beliefs and attitudes. So do the composition of novels and other kinds of creative writing. So too do history and sociology and literary studies. Further kinds of intelligence again are embodied in musical composition and performance, in art and art appreciation, and, differently, in philosophy. Several of these fields open up new possibilities to humanity, some, though not all, through what could be called 'generating and testing hypotheses'. Hence Dennett's tower is a truncated one. Mary Midgley has rightly criticized the view

25 Dennett, ibid., p. 381.

26 Dennett, ibid., pp. 389–93.

which re-surfaces here that science is the supreme role of the mind and thus the destiny of evolution.[27]

Besides, the controversial phenomenon of consciousness fails to figure in this story, although Dennett has written about it elsewhere.[28] This may be because Dennett's account is 'an outrageously oversimplified structure',[29] but some important questions about its role are raised nevertheless; and whether we adopt a reductionist or a non-reductionist theory of consciousness is fortunately beside the point for present purposes, which require no more than the ordinary, familiar concept of consciousness. Reinforcement (as in Skinnerian creatures) may not presuppose consciousness, and the soliloquies with which Dennett endows these creatures cannot be taken literally. But Popperian creatures (such as sea-gulls) seem to require consciousness, in order to be able to perceive opportunities and dangers, and to entertain the hypotheses that they discard (unless this entertaining of hypotheses is also a metaphor, in which case it throws no clear light on our own Popperian moments and tendencies). Without consciousness, creatures would seem to lack control over behavioural options. Amphibians, reptiles and birds must have it as well as mammals (for example, when they notice prey or predators), whether or not invertebrates do (and at least octopuses, said by Dennett to be truly smart,[30] seem to be Popperian creatures). But in the absence of consciousness from mere Darwinian creatures, they can only metaphorically be regarded as 'trial and error learners',[31] for trial is a deliberate activity, and not just serendipitous or random gyration; indeed pre-Popperian creatures can hardly be held to be 'learners' at all. Human trial-and-error procedures are typically linguistic, and nothing less than Gregorian in their dependence on devices such as laboratory apparatus and telephone directories.

Furthermore, the entertaining soliloqies that Dennett ascribes to Skinnerian and Popperian creatures make no sense in the absence of a concept of the self; for without such a concept, these creatures cannot ask 'What should *I* think about next?', as Dennett suggests (but the italics are mine).[32] By contrast, the subset of Gregorian creatures who can 'think better about what they should think about next'[33] must have some kind of rudimentary concept of the self. But this is also the subset comprising linguistic creatures; and it is language (systems of signs structured by syntax) that makes possible the kind of generalizing (with sortal

---

27 Mary Midgley (1985), *Evolution as a Religion*, p. 73.

28 Dennett (1991), *Consciousness Explained.*

29 Dennett, *Darwin's Dangerous Idea*, p. 373.

30 Dennett, ibid., p. 376.

31 Dennett, ibid., p. 369.

32 Dennett, ibid., p. 378. In Dennett (2003), *Freedom Evolves*, Dennett recognizes at p. 249 that the concept of the self is lacked by the run of Popperian creatures, suggesting that it is made possible by language.

33 Dennett, *Darwin's Dangerous Idea*, p. 378.

predicates[34] such as 'human being' used of subjects) and individuating (with names, pronouns and demonstratives employed as referring expressions) required for thought about 'me' and 'you'. Certainly there is evidence of non-linguistic creatures having a concept of the self (recognizing reflections of themselves in mirrors and rivers); but *reflection on the self* requires a system of reference for which language is probably indispensable. Further, the subset of Gregorian creatures who are capable of second-order beliefs about the beliefs and attitudes of others would require a fairly developed concept of themselves as one among others, and thus language to make this possible. Thus the kinds of understanding involving either thinking about thinking or thinking about the thoughts of others are probably open to linguistic beings only. Little light is thrown on these kinds of understanding by reflection on Popperian or Gregorian creatures as such.

Admittedly, tool-use (a possibility applicable to Gregorian creatures in general) may help explain the early stages of acquiring the concept of the self, since with tool-use part of the environment becomes an extension of one's body, and encourages the wielder of tools to distinguish its extended body from the rest of the environment. But tool-use involves culture; there are right and wrong ways of using a stick to fish for termites, just as there are for using hammers (and words), and these skills and conventions result from conscious discoveries. Culture may not invariably require language (despite Dennett's claim to the contrary[35]), but unless Dennett can persuade us that culture is generated by an invasion of memes (see below, and Chapter Six, Section 1 above),[36] it is difficult to see how either its origins or its development are to be explained by Darwinian (or analogous) processes. This is precisely where conscious selection becomes relevant, as opposed to natural selection. (And if nothing is allowed to contrast with natural selection, then 'natural selection' loses its meaning.)

There is also a question about the aptness of regarding words as tools. Tools are devices for performing purposes that can be specified independently; yet many linguistic activities cannot be specified in the absence of the segments of language that constitute them, such as greeting, cursing, apologizing and thanking. If so, then words are in a different league from tools. Certainly there are pre-linguistic practices that could have evolved into (say) greeting. But it is hardly plausible that the linguistic practice of greeting arose through Darwinian processes, whether of genetic or of memetic evolution.

There again, whole new dimensions of intelligence, meaning and creativity are generated by the capacity to conceive of the future and the past, and of possible worlds. We can now conceive of ourselves as continuing entities for whose future we can form plans. We can foresee our own deaths. We can imagine alternative outcomes of present action. And we can reflect on the causes of events, and

---

34 For a fuller exposition of sortal predicates, see P.F. Strawson (1959), *Individuals*, 1959, pp. 168–73.

35 Dennett, *Darwin's Dangerous Idea*, p. 341.

36 Dennett, ibid., pp. 341–2.

on responsibility for actions and outcomes. Simpler forms of language are to be found which fail to provide for these possibilities. So acquiring language with tenses and conditionals and modal auxiliaries constitutes several huge accomplishments. Sooner than deny any of this, Dennett would ascribe these steps to an uncontrollable invasion of self-replicating memes (for futurity, conditionality, etc.). But these so-called 'memes' amount in fact to the culturally transmitted discoveries of conscious linguistic creatures. As Dennett says, there is no doubt that culture evolves, in a Darwin-neutral sense of that term, that is, undergoes modification over time.[37] But the view that the process by which this takes place is analogous to, let alone identical with, Darwinian selection ignores, or fails sufficiently to acknowledge, the rationality and the inventive and autonomous creativity of those who participate in the selection of ideas, endorse some and reject others for reasons that seem good ones, and often contribute original ones not randomly but because they consider them improvements. Thus Gould's view is to be preferred that cultural evolution operates in an importantly different manner from biological evolution.[38]

What implications are there for the evolution of meaning? Natural selection, as was argued above in Chapter Eight, cannot explain progress in science, and the same applies to other cultural achievements, and, I want to suggest, to meaningful action in general; indeed, this can be explained by natural selection no more than scientific progress can. For meaningful action (on which more will be said. in the coming section), like contributions to disciplines such as science, involves self-conscious (and often rational) activities, discretion, and selection of actions against a background of criteria and priorities. While Dennett's Tower explains some of the prerequisites of meaningful action, and similarly of creativity, such as capacities for the discarding of some hypotheses and the adoption of others, it does not suffice to explain meaningful action itself, or creativity in science, art or the arts. Indeed, if meaningfulness (whether of language or of action) is, as David Cooper contends, conceptually inseparable from the shared practices of human existence, then the evolution of human beings from (pre-cultural) nature to culture (or second nature) cannot be explained by Darwinian evolutionary processes involving earlier levels or layers of meaning; for meaning (in the sense of meaningful action) is not to be found apart from these shared human practices.[39] Meaning might still evolve (in the sense of 'emerge') from long processes of social interaction; but importantly it could not evolve in the sense of undergoing successive modifications of a Darwinian kind on the way to this emergence.

Before we turn, as promised, to meaningful action, some possible implications for freedom should be brought to attention, however summarily. Dennett's explanations of the eventual emergence of cultural activities like science (such as

---

[37] Dennett, ibid., p. 345.

[38] Stephen Jay Gould (1991), *Bully for Brontosaurus*, p. 63.

[39] David E. Cooper (2003), *Meaning*, p. 38. See also Stephen .R.L. Clark (2000), 'The Evolution of Language: Truth and Lies', *Philosophy*, **75**, 401–21.

his appeal to Popperian processes of pre-selection and de-selection), insightful as they sometimes are, may not be as deterministic as he claims. For pre-selection must be considered a variety of selection, and there is at least a prima facie case for holding that Popperian creatures and their subsets and successors have a rudimentary capacity to select among alternative courses of action, and that they often could have done otherwise. While this would not involve fully fledged libertarian freedom in the absence of linguistic capacities and of the possibility of conscious reflection on alternatives, there would even so be some kind of conscious selection between behavioural options presented to these creatures by their perception of the environment. So Dennett's narrative, contrary to his intentions, may portray early stages of the evolution (again in a Darwin-neutral sense) of libertarian freedom. Dennett would reject this view, for a deterministic account of freedom and its antecedents is supplied in his more recent *Freedom Evolves*.[40] But it remains unclear that the notion of pre-selection (as ascribed to Popperian creatures) is available for such a deterministic account, at least if it is to help explain the capacities of Gregorian and of linguistic creatures for selection (in the standard, arguably non-deterministic sense) in the world of culture.

If so, Darwinian processes, construed as deterministic processes, will perforce be unable to explain the emergence either of creativity or of culture, or therefore of meaningful communication and action. However, it is far from obvious that Darwinian processes need be regarded as deterministic at all (or as 'algorithmic', to use Dennett's favourite expression in *Darwin's Dangerous Idea*). Once we are liberated from having to regard the processes of Dennett's Tower (or likewise Darwinism itself) as deterministic, we can far more readily discern in these processes (such as the emergence of pre-selection and of Popperian creatures from Skinnerian ones) some of the origins of intelligence, understanding and intentionality, or at least of the emergence of their prerequisites; for it is now more readily possible to recognize some kind of continuity between pre-human and human intelligence. Thus Dennett's processes can at least be seen as providing the springboard from which the practices of human life and culture could be launched, or could launch themselves, and with them meaningful communication and action. Springboards are neither unnatural makers of miracles (as Dennett considers skyhooks to be: but I am not hinting at supernatural interventions), nor are their products predictable (like the constructions generated on building sites by cranes), but enable whole repertoires of leaps to be learned and dives to be discovered.[41] So it is, I suggest, with the genesis of meaning and culture.

---

40 Dennett, *Freedom Evolves*.

41 For circumstances in which the leap into language could have been leapt, see Clark, 'The Evolution of Language'. Clark sensibly resists Steven Pinker's view (see Pinker (1994), *The Language Instinct*, pp. 362–4) that, even though we have no idea how this could have happened, language must have been generated by natural selection. For unnatural, inventive human selection cannot be ruled out.

This stance, it should be added, is consistent with conclusions adopted above. For this stance, unlike Dennett's, can be reconciled with a view of evolution as combining laws of nature with provision for libertarian freedom (as also advocated by Ward: see Chapter Eight above). It coheres, again, with the deployment above in Chapter Seven of a version of the Free Will Defence with regard to moral evils and atrocities, something not open to any deterministic account of human powers. Finally, unlike forms of Darwinism that deny progress and purposiveness in evolution, it also coheres with belief in evolution having an overall purpose, in the form of the generation of intrinsic value, both in human and in non-human life. Dennett, while denying purposiveness, seems not to reject progress in the form of sophistication of design-structures, with science as their pinnacle. But, as argued above, that particular version of belief in evolutionary progress need not be endorsed even if the entire process is interpreted as purposive and enabling progress in the form of the emergence of intrinsic value.

Section 5: Meaningful Action, Meaningful Work and a Meaningful Life

In this section, meaningful action is related to meaningful work and to a meaningful life. As was mentioned in Section 2 above, utterances are distinctively significant or meaningful ones when they resonate with the central priorities or the deeper concerns of speaker or audience. (A wider range of speech and writing is meaningful in the related, broader sense of having coherence, being fit, that is, for communication, whether to self or to others, and thus avoiding meaninglessness; but here 'significant' is used in the more demanding sense just presented.) Actions too can be meaningful through bearing a parallel relation to priorities or concerns (and so indeed can lives); for the agents of meaningful actions (and lives) are able to identify with their deeds and roles, and sometimes to regard them as having significance in a larger scheme or in the shared practices of communal life. There is more to a meaningful life than this, for perceptions of the universe or of human history (for example as pervaded by purposiveness) can enhance awareness of life's meaning. Yet a life can be meaningful simply through being understood as playing a contributory role in a community.

This point epitomizes John Cottingham's claim that 'meaningfulness is what might be called a hermeneutic concept: 'for something to be meaningful to an agent, that agent must *interpret* or *construe* it in a certain way' (Cottingham's italics).[42] My qualification is that, since the interpretation can take place long after the action thus interpreted, but the action (first meeting a future spouse, say) can still be held to have been meaningful all along, and thus before it was regarded as meaningful, the requirement for something to be a meaningful action is that it be capable of such interpretation or construal (as related to some profound or

---

[42] Cottingham, op. cit., pp. 21–2. 'Hermeneutics' is further explained in Kevin J. Vanhoozer (1998), *'Is There a Meaning in This Text?'*, pp. 22–3.

serious human purpose, to supply Cottingham's own account). The hermeneutic aspect of the concept of meaningfulness links it, as Cottingham remarks, with 'its semantic home within the domain of language',[43] and is relevant to the meaningfulness of a life as well as to that of an action.

Accordingly, as was remarked above in Chapter One, both meaningful action and a meaningful life may be held to involve the holding by the relevant agent of a set of more or less integrated priorities, and meaningful communication may be held to derive its significance, in part, from and against this background. The requirements for holding integrated priorities are unsurprisingly stronger for a meaningful life than for meaningful action. Meaningful action and communication could even derive some of their significance from rejection of communal norms (as when Luther pinned his Ninety-five Theses to the church door at Wittenberg, rejecting some shared values, but implicitly endorsing others through his very act of communication). By contrast, as Cottingham has written about lives that are meaningful, 'the meaningful life for human beings is an *integrated* life – one where my pet projects and plans are not kept in an isolated category which allows me to pursue them perpetually shielded from the demands on me as a parent, or a friend, or a colleague, or a citizen'.[44] The compartmentalized life is not, as Cottingham observes, a fully human life, nor, for that reason, a meaningful one.

With meaningful action, however, the important background requirement consists in being clear about one's own purposes and the relation to them of the current step: as Cottingham puts it, 'meaningfulness in action implies a certain degree of *self-awareness* or *transparency to the agent*; for me to engage in meaningful activity I must have some grasp of what I am doing, and my interpretation of it must reflect purposes of my own that are more or less transparent to me'.[45] This requirement of self-awareness may throw light on Hook's remark, cited in Section 2, that humans are uniquely situated to make the value that permeates the biosphere decisively meaningful.[46] For to make such value meaningful would involve adopting a self-conscious role in which the value distributed across the biosphere is related to the pursuit of shared, conscious purposes through meaningful activity. (Hook may be going further, and using 'meaningful' in a sense related to his near-identification of meaning and value; but grounds were given above for being sceptical about that position.) To this possibility I will be returning; here it can be remarked that what might make humans 'uniquely situated' in this regard is, in part, the very capacity for self-awareness and its relation to action remarked by Cottingham.

In a section entitled 'The Meaning of Work', John White has put forward an account of meaningful work that may seem to comply with the account of

---

43 Cottingham, op. cit., p. 21.

44 Cottingham, ibid., p. 29.

45 Cottingham, ibid., p. 22.

46 Hook, op. cit., p. 372.

meaningful action just given. According to White, meaningful work or its end-product must figure among an agent's major goals in life, and its value turns entirely on fulfilment of the agent's autonomy.[47] My own account, of which White is critical, is that meaningful work is free and creative productive activity (whether productive of commodities or services or theories or works of art) where workers care about standards of production, exercise skill and judgement and have a say in working arrangements, without their major goals in life necessarily being involved. Such work could be one of many components of a year's activity, or could be temporary work, or a regular activity not crucial to the worker's sense of identity, without ceasing to be meaningful.[48] The meaningfulness of such work derives from the worker's concern for shared standards of workmanship and autonomous contributions (skill, judgement, contributing to decisions) to a shared process; for the intrinsic value of such work need not derive (entirely or even partially) from the agent's autonomous choices (although the agent's autonomy could still play a part), but would typically derive from exercise of a range of essential human capacities.

It might appear that my endorsement of Cottingham's account of meaningful action (in terms of awareness of one's purposes, and of the relation of current activity to them) commits me to something close to White's account of meaningful work, since what goes for meaningful action should apply equally to meaningful work. However, while work is obviously a kind of activity, its creative and productive nature introduces additional standards of meaningfulness (relating to workmanship), which bring into play human potentials not invariably involved in action as such. Related shared standards of creative and productive excellence supply work, as opposed to action, with ways in which workers can interpret and take pride in their contribution, even when the product is not among the agent's major goals. Cottingham would probably not endorse White's account either, but for additional reasons. For his self-awareness or transparency condition of meaningful action is not satisfied in every case of autonomous action. Cases in which agents are unaware that their goals have become obsessive are recognized by Cottingham to be counter-instances.[49] Further, for a life to be meaningful, it is insufficient for actions to be performed for personal satisfaction; they have to be capable of being informed, he holds, 'by a sense of the worthwhile part they

---

47 John White (1997), *Education and the End of Work*, pp. 4–7.

48 Robin Attfield (1974), 'On Being Human', *Inquiry*, **17**, 175–92; Attfield (1984), 'Work and the Human Essence', *Journal of Applied Philosophy*, **1**.1, 141–50; Attfield (1995), *Value, Obligation and Meta-Ethics*, Chapter 4. A reply to White can be found in Attfield (2001), 'Meaningful Work and Full Employment', *Reason in Practice*, **1**.1, 41–8. Curiously, a similar view can be found in the thought of the logical positivist, Moritz Schlick (see Oswald Hanfling (1987), *The Quest for Meaning*, p. 37), who located the meaning of life in 'the joy of sheer creation' and in playfulness.

49 Cottingham, op. cit., pp. 22–3.

play in the growth and flowering of each unique human individual'.[50] While this condition may be couched in over-demanding terms, it well conveys that the agent needs to regard their actions as, in general, objectively worthwhile (rather than personally beneficial), a requirement which makes good sense too for meaningful work, and coheres well with the account of meaningful work presented above.

Both the previous point about value and Cottingham's transparency and worthwhileness requirements bear out the contention that autonomous choice is not the sole source either of value or of meaningfulness.[51] My account of meaningful work, if accepted, also supplements the above account of meaningful action by adding another sufficient condition; for meaningful work that might not have counted as meaningful action on that account (perhaps through lack of connection with major life-goals) now turns out to be a form of action that is recognizably meaningful on the basis just presented (skill, judgement, shared standards, etc.). Such work may or may not form part of a meaningful life, since for that there are stronger requirements (see above and below); but the same holds good of meaningful action. Nevertheless, meaningful work, like most other forms of meaningful action, can help a life to be or become meaningful through supplying a basis for self-respect, and for the sense of compliance with worthwhile standards that goes with it,[52] and through making agents recognize that at least some of their activities contribute to shared goals.

Section 6: Darwinism and Meaningful Life

Cottingham adds that the journey of a truly meaningful life, far from being facilitated by just any set of purposes or projects selected by the traveller of that journey, cannot be made without regard for the conditions of our flourishing as human beings. A worthwhile life will embody value linked to the pursuit of what is objectively conducive to the flourishing of human nature. Cottingham recognizes that such journeys could take multiple forms, theistic forms comprising just one kind, and does not in any case say that life cannot be meaningful at all, as opposed to 'truly meaningful', in the absence of such a pursuit.[53] But he is concerned that current scientific orthodoxy may either seem to undermine the possibility of meaningful lives, or greatly reduce their prospects. For he suggests that it seems to imply that 'our entire human existence is not much more than a random blip on the face of the cosmos',[54] and thus valueless. Nor is he alone in harbouring such concern: Mary Midgley, replying recently to Dawkins, remarks

---

50 Cottingham, ibid., p. 31. See also p. 32.

51 See further Attfield, *Value, Obligation and Meta-Ethics*, pp. 63–5.

52 For the relevance and intrinsic value of self-respect, see Attfield, 'Work and the Human Essence', pp. 144–5, *Value, Obligation and Meta-Ethics*, pp. 65–6, and 'Meaningful Work and Full Employment', pp. 46–8.

53 Cottingham, op. cit., pp. 32–3.

54 Cottingham, ibid., p. 31.

that the neo-Darwinist orthodoxy that maintains that 'natural selection is the sole and exclusive cause of evolution' makes 'the world, therefore, in some important sense, entirely random'.[55]

Cottingham's positive claims about meaningful lives cohere well with the stance adopted here, and with a fuller account of worthwhile lives that I have presented elsewhere.[56] One qualification should be made; for a life could, I suggest, be understood as meaningful that focussed on non-human flourishing (plus the well-being of the human agent concerned), rather than on human flourishing, or on the latter exclusively. But what of Cottingham's negative comment on the apparent implications of scientific orthodoxy? While this could seem to be borne out by Atkins's remarks in *The Creation* about the insignificance of human life,[57] Cottingham undoubtedly has in mind biology as well as physics, and his claim is most relevant here insofar as it concerns Darwinism (which he later concludes to be cotenable with theism). So it will be considered on that basis.

Need Darwinism undermine the prospects of its adherents for meaningful lives, and for belief in the kind of value that goes with life being worth living? For 'Mind-First' Darwinism, no such implication arises, for both the laws and the mutations that facilitate evolution and its valuable products are generated by a cosmic designer in whose purposes for humanity, society and nature meaningful lives can be found. (One kind of example will be offered in the coming section and the coming chapter.) Again, the so-called Blank-Paper variety of Darwinism (to stay with Richards's nomenclature) has declared itself through Stephen Jay Gould (and his NOMA principle, discussed above in Chapter Six) as compatible with religion and ethics, and distinguishes between the separate and compatible spheres of science and of meaning. For such an approach, despite Darwin's fears that suffering might be meaningless,[58] there is no ultimate incompatibility between Darwinism and any of the various shared visions widely recognized as making life meaningful, religious and humanistic ones included.[59] The problem arises, if at all, with the Gene-Machine version of Darwinism, which sometimes represents human lives as merely the vehicles of the genes and memes by which they are moulded and for the sake of whose survival they exist; as Midgley suggests, less deterministic approaches, or ones recognizing factors other than natural selection, need have no such implications.

But are there no prospects for the lives of consistent adherents of the Gene-Machine approach to be meaningful? Its central adherents claim otherwise. For Dawkins actually denies that we are compelled to obey our genes, remarking that 'We alone, on earth, can rebel against the tyranny of the selfish replicators';[60]

---

55 Mary Midgley (2005), letter to *The Guardian*, 6 September, p. 23.

56 Robin Attfield, *Value, Obligation and Meta-Ethics*, pp. 63–78.

57 Peter Atkins (1981), *The Creation*, p. 3.

58 See Chapter Six, Section 4 (above).

59 Gould (1999), *Rocks of Ages*, p. 52.

60 Richard Dawkins (1976), *The Selfish Gene*, p. 205.

and this qualification to a mechanistic view of human life (if consistently carried through so as genuinely to facilitate the capacity for meaningful rebellion) might allow scope for meaningful participation in activities such as science. Dennett also praises the pursuit of science as the summit of his Tower, recognizes the wonder and the high value of the products of design (both natural and anthropogenic),[61] and claims that control of our lives, and thus the ability to shape them with such values in mind, is compatible with deterministic Darwinism.[62] While this is a dubious claim, we can see how people who stress these particular planks of Dennett's platform (rather than others) could find scope to interpret their lives as meaningful, since meaning is, as Cottingham remarks, a hermeneutic concept. For such lives to count as 'truly meaningful' in Cottingham's sense, the adoption of objective values by the people concerned would, if Cottingham is right, also be required, but that too cannot be excluded if Dennett's evolutionary account of ethics is in any degree successful,[63] an issue which cannot be discussed further here, or if objective values are recognized on some other basis. E.O. Wilson too believes that evolution itself generates value and conveys meaning to the lives of its human adherents.[64] Thus, while the Gene-Machine (or Neo-Darwinist) approach could well undermine belief in life's meaningfulness, vindicating the concerns of Cottingham and Midgley, it may be possible to construe (or to qualify) this approach in such a way that prospects remain for the lives of its adherents to be meaningful. The danger is that its adherents are sometimes prone to construe it otherwise.

It now remains to characterize one particular kind of role in which agents could regard their action or work as contributing to or protecting value, and thus as worthwhile through fostering goals both individual and shared. Given sufficient self-awareness, this would then count as meaningful action. It could also give substance to Hook's suggestion about the human ability to make the value spread out in nature 'decisively meaningful'. This is the subject of the final section of the present chapter.

Section 7: Stewardship

When we consider roles appropriate to humanity in relation to nature, several roles emerge as proper and defensible. Supplying one's basic needs is clearly indispensable for survival, and any attempt to disown this role would ignore the basis on which humanity has evolved. Making a living is no more than a modern equivalent. A related role is that of making nature habitable, a role required for

---

61 Dennett, *Darwin's Dangerous Idea*, pp. 380, 520.

62 Dennett, *Freedom Evolves*, pp. 281–4.

63 Dennett, *Darwin's Dangerous Idea*, pp. 453–510.

64 Wilson's stance is discussed in Chapter Ten, Section 4 below. See also Michael Ruse (2001), *Can a Darwinian?*, p. 183.

sustainable, intergenerational survival. This does not require transforming the entire surface of the planet, but does involve modifying parts of the environment, rather than letting it remain a wilderness, so as to make possible shelter from the elements, and homes for human communities. All these activities can in the right circumstances contribute to shared values and to living a meaningful life, particularly when conducted with sensitivity to the needs of nature.

Further roles involve a greater awareness of nature's value, whether of its instrumental value for humans or of its intrinsic value. One is that of conserving natural resources, including renewable resources such as forests and fish-stocks, as well as non-renewable mineral resources. If renewable resources are to be conserved for future generations, some understanding must be shown of the limits beyond which this would be impossible, and thus of the needs of the relevant living creatures; and if they are being conserved not only for future consumption but also for future study or appreciation, then a greater awareness of their good becomes indispensable. This introduces the overlapping role of the preservation of species, sub-species and habitats, either for the sake of future study or appreciation, or for the sake of the creatures themselves, and their own future generations. Preservation can take the form of non-interference and letting-be, but can also assume more active forms, adopted to preserve vulnerable species or ecosystems.[65] Another of the more active forms comprises rehabilitating nature in areas that have become degraded, so as to restore a habitat or ecosystem to something like its condition prior to previous exploitation (which is unlikely to be any kind of primeval condition, since nature has been constantly changing since life first began evolving, and before that too).[66]

Conservation and preservation can both be subsumed within the possible human role of *stewardship* of planetary nature, and of the realm of space that human activity can affect. The various religious and secular forms that stewardship can adopt will be discussed further in the coming chapter, as will some of the more central criticisms to which it is sometimes subjected. But it should be remarked at once that the idea has religious origins in the theistic belief that humanity is God's steward, deputy or agent, and entrusted and answerable as such with the care or the management of goods such as wealth or time or (as in the current case) the lands, seas and ecosystems of the planet. This belief was classically expressed by the seventeenth-century Chief Justice, Sir Matthew Hale:

> The end of man's creation was, that he should be the viceroy of the great God of heaven and earth in this inferior world; his steward, *villicus*, bailiff or farmer of this goodly farm of the lower world … invested with power, authority, right, dominion, trust and care, to correct and abridge the excesses of the fiercer animals, to give protection and

---

65 Conservation and preservation have often been contrasted, sometimes on an erroneous basis. See Bryan G. Norton (1986), 'Conservation and Preservation: A Conceptual Rehabilitation', *Environmental Ethics*, **8**.3, 194–220.

66 See further William Throop (ed.) (2000), *Environmental Restoration*.

> defence to the mansuete and useful, to preserve the species of diverse vegetables, ... [and] to preserve the face of the earth in beauty, usefulness and fruitfulness.[67]

Whether or not stewardship is a central purpose of human existence, it has clearly been held to involve answerability as well as responsibility. However, answerability could be regarded as owed to humanity, or the intergenerational human community, as well as or instead of to God, and in this way stewardship can be held on a secular basis.

The above mention of management could be thought to suggest that stewardship is essentially an instrumentalist approach, like that of a farm-manager trained in a business school, aiming to maximize profits (over a long-term period) for the farm's owner, if necessary by selling off the less productive holdings of the farm. But stewardship is not to be confused (despite what some of its critics say) with managerialism,[68] even in versions which are entirely anthropocentric, let alone in non-anthropocentric versions like that of the philosopher Joel Feinberg, who wrote of 'our housekeeping role, as temporary inhabitants of this planet'.[69] Alongside management, the goal of care was also included above, involving concern for the good of the creatures and habitats concerned (whether or not for the sake of long-term human benefit); and this stance precludes a managerialist approach, and involves a recognition of the high value (whether instrumental, inherent or intrinsic) of the objects of care. Indeed another near-synonym for stewardship (besides management) is 'trusteeship', a practice in which responsible care is taken on behalf of others, usually to preserve some good over a long period. It is also sometimes suggested that this kind of stewardship resembles being the curator of treasures, which are to be preserved because of their inherent value; this may present too static a picture of stewardship, but is at least as appropriate as the model of management. So is the further model of guardian, use of which has the advantage of recognizing the possibility of dynamic and autonomous development on the part of whatever the guardian holds in trust.[70] Relatedly, stewardship need not be understood as having an anthropocentric grounding,

---

[67] This passage of Hale is quoted in John Passmore (1974), *Man's Responsibility for Nature*, p. 30, and is here abbreviated.

[68] For such criticism, see Richard Routley and Val Routley (1980), 'Human Chauvinism and Environmental Ethics', in Don Mannison, Michael McRobbie and Richard Routley (eds), *Environmental Philosophy*, 96–189; Richard J. Evanoff (2005), 'Reconciling Realism and Constructivism in Environmental Ethics', *Environmental Values*, **14**.1, 61–81, p. 75.

[69] Joel Feinberg (1974), 'The Rights of Animals and Unborn Generations', in William T. Blackstone (ed.), *Philosophy and Environmental Crisis*, 43–68, p. 56.

[70] The roles of trustees, of curators and of guardians are cited as near-equivalents of that of stewards in a statement of the General Synod of the Church of England Board for Social Responsibility (1991), *Christians and the Environment*, p. 2. See also Attfield (1999), *The Ethics of the Global Environment*, pp. 47–8. For the terms 'instrumental', 'inherent' and 'intrinsic value', see Chapter Eight, Section 2 above.

despite anthropocentric exponents such as Calvin, and can be (and historically has often been) understood as recognizing the intrinsic value of the creatures over which it is exercised.

On most accounts, stewardship covers both conservation and preservation (sometimes of the letting-be variety), and can embrace rehabilitation as well. It can thus recognize the intrinsic value both of human flourishing, the usual justification of conservation, and sometimes the goal of preservation and rehabilitation, and also of the flourishing of non-human creatures, animal and vegetable, which is often recognized as part of the reason for preservation (and for rehabilitation too). Its participants can thus see their activity as promoting shared, objective values, and as a contribution towards performing the role of answerable agents, whether answerable to God, to humanity or to both. Given also sufficient self-awareness, it can be recognized as meaningful action, and potentially part of a meaningful life (and perhaps also as a human enactment of God's creative purpose, as manifested in evolution). Many forms of human activity, including gardening and farming, fishing and forestry, as well as environmental campaigning, can be understood as meaningful on such a basis.

This does not make the role of stewardship distinctively meaningful, since the other roles mentioned earlier can also be so understood in appropriate circumstances, as well as many others. What confers its distinctiveness is its promotion or preservation of nature's unrivalled value, and at the same time its care of the whole range of the magnificent and varied products of evolution. How best to care for the products of evolution will be briefly discussed in Chapter Ten below, but stewardship unquestionably involves such care. The value of the biodiversity that it seeks to conserve may be indirect (see above, Chapter Eight, Section 3), but remains crucial to the future both of humanity and of other species.[71] Indeed, as Peter G. Brown claims, stewardship of the environment 'extends the principle of equal respect through time',[72] as long as we bear in mind that what are to be respected equally are equal interests, and not necessarily human interests alone. But human language, self-awareness and conscious self-direction (what is sometimes called 'self-creation') make this role distinctively available to humanity, and thus make the conservation and preservation of nature meaningful activities of a distinctive kind, since (as Hook suggests) the distinctive value that is already present in nature is here recognized, read and communicated, and in these ways made meaningful.

In the course of discussing meaning in this chapter I have considered why it is no coincidence that we speak both of the meaning of expressions and of the meaning of actions and of life. Dennett's narrative of the evolutionary development of design and intelligence has been found to throw light on our capacities for meaning and understanding (and thus the relation between

---

[71] This is underlined in the final chapter of Dennett's *Darwin's Dangerous Idea*, in a section entitled 'In Praise of Biodiversity' (pp. 511–20).

[72] Peter G. Brown (2000), *Ethics, Economics and International Relations*, p. 30.

evolution and meaning), without vindicating either Darwinian or deterministic explanations of meaning or culture. Meaning turns out to be grounded in shared human practices, and meaningful actions (and lives) turn out to derive their meaning from transparent, integrated sets of priorities, while meaningfulness can be enhanced by a broad, sometimes cosmic vision of life or history. Darwinism need not undermine such meaningfulness in any of its versions, although Gene-Machine Darwinism can appear to put it at risk. One such cosmic vision of life can be found (at least in part) when we understand ourselves in the role of steward, trustee or guardian of planetary nature, and thus of the valuable fruits or products of evolution.

Chapter Ten

The Ethics and Metaphysics of Stewardship

Section 1: Introduction

In this final chapter, the ethics and metaphysics of stewardship are considered, together with some of its presuppositions. The need for an ethic corresponding to the sphere of human influence and to the kind of value-theory introduced above is related both to criticisms and to defences of the stewardship approach. Ethical relations between successive generations are also related to stewardship and its responsibilities. The metaphysics of both secular and theistic versions of stewardship are then discussed, and related to ways in which the practice of stewardship can be motivated and life can be found meaningful. Theistic stewardship turns out to be motivationally more self-sustaining simply because of its distinctive metaphysic; this is, of course, the same metaphysic that previous chapters have argued to be well supported by some of the theistic arguments.

A section on the experience of givenness, and its strong relation to stewardship and to E.O. Wilson's concept of biophilia, concludes the discussion of stewardship, and leads into the final section in which the overall argument of the book is summarized, and diverse strands linking creation, evolution and meaning are brought together, including (not least) the contributions of Darwinism itself to arguments for the consistency and rationality of belief in creation.

Section 2: Ethics: Scope, Method and Content

The growth of human numbers and, more particularly, of human technology has vastly increased the scope of human impacts in time and in space. These impacts now affect our entire planet and its environs, most if not all of its species, and most if not all future generations. Many ethical systems, however, have been principally (if not exclusively) concerned with person-to-person relations in the present. As Hans Jonas has argued, we need an ethic the scope of which corresponds to the sphere of these impacts,[1] for otherwise responsibilities will be recognized for an unduly limited sphere, and a great deal of good and harm generated by human action would be made to seem to fall outside the scope of

---

1 Hans Jonas (1984), *The Imperative of Responsibility*.

ethics. But it is unnecessary to do more than make these implications explicit to uphold Jonas's conclusion; for an ethic outside whose scope fell vast swathes of humanly generated good and harm would be manifestly inadequate, in virtue of the very concepts of ethics, good, harm and morality. Whatever we hold about ethical method, these concepts imply, in conjunction with the facts about human impacts, that the scope of ethical responsibilities extends to the full range of these impacts wherever it would be possible to foresee them at the time for action (or inaction).

The same conclusion tallies with the value-theory introduced above, and is supported if that theory is endorsed. For that theory, besides locating intrinsic value in the development and exercise of essential human capacities, recognized it also in the flourishing of living creatures in general (and thus in the development of their essential capacities): see Chapter Eight, Section 2 above, where thought-experiments and related grounds are adduced in support. But the existence (actual or possible) of intrinsic value necessarily supplies a reason for action for rational agents capable of promoting, preserving or cherishing that value. So there are reasons for action wherever human impacts can make a difference to intrinsic value across space and time. And where the difference is non-trivial and the balance of reasons supports one rather than another course of action (or inaction) there will be a corresponding responsibility.[2] The method adopted in this reasoning turns on there being a necessary connection between intrinsic value, reasons for action and responsibility; and that there is such a necessary connection has been cogently argued by Kenneth Goodpaster.[3] This method also coheres well with the biocentric consequentialist normative ethic that I have defended elsewhere;[4] but there is no need to defend it here, granted that adherents of stewardship need neither be biocentrists nor consequentialists.

Belief in human stewardship of the planet and its environs firmly places responsibility for human impacts on human shoulders, while recognizing also how responsibility can be limited by lack of knowledge, resources or power, and is fully consistent with Jonas's conclusions, and endorsed by Jonas himself.[5] This recognition of human responsibility for the biosphere is sometimes made a matter for complaint. Thus Richard J. Evanoff criticizes the stewardship model for authorizing activist human management of nature in general, despite there always being more to nature than human agents can understand.[6] Evanoff would be right if he had in mind there being spatial limits to stewardship (since

---

[2] See further Robin Attfield (1995), *Value, Obligation and Meta-Ethics*, Chapters 3 to 8.

[3] Kenneth Goodpaster (1978), 'On Being Morally Considerable'.

[4] In Attfield, *Value, Obligation and Meta-Ethics*, Attfield (1999), *The Ethics of the Global Environment*, and Attfield (2003), *Environmental Ethics*.

[5] Jonas, op. cit., p. 8.

[6] Richard J. Evanoff (2005), 'Reconciling Realism and Constructivism in Environmental Ethics', p. 75.

the galaxies largely lie beyond human influence), and also about the need to avoid policies based on ignorance, and risking irreversible change to nature here on Earth. But stewardship is compatible with unswerving observance of the Precautionary Principle where irreversible harms or the crossing of ecological thresholds are in question.[7] Besides, as was mentioned in Chapter Nine, Section 7 above, stewardship does not involve managerialism; it can be grounded in recognition of the value of non-human creatures, and is compatible with leaving wilderness, habitats and species well alone (an approach that Evanoff actually contrasts with stewardship). The philosopher Willem B. Drees has recently written supportively of such letting-be (which he calls 'reticence', and contrasts with 'actively changing nature') as characteristic of stewardship. (He also regards stewardship as 'taking up the work God entrusted to us', a theme to be revisited in the coming section.[8]) Adherents of stewardship, as he suggests, will often find non-intervention the most consistent way of performing their role. But granted the expanded sphere of human impacts, it is in any case crucial that responsibility for such impacts be recognized as borne by the individuals, corporations and countries that generate them; for the alternative is power without responsibility, or rather without its recognition.

Here it might be objected that our evolutionary inheritance makes responsible action on such a scale impossible, and therefore not a responsibility. For this inheritance is sometimes held by sociobiologists to restrict our behaviour to actions favouring our kin, plus beneficent actions to others that are likely to be reciprocated; and behaviour that benefits other species and future generations unrelated to ourselves well exceeds this range. An empirical response to this objection is that many people do actually take trouble and make sacrifices to relieve animal suffering, or campaign to curtail global warming largely for the sake of future generations; and as we have seen, even Richard Dawkins, author of *The Selfish Gene*, holds that humanity is capable of rebelling against the imperatives of the genetic forces by which he believes we are largely governed.[9] For his part, Peter Singer has supplied an attractive theory that both counters the objection and helps explain the extremely wide range of human behaviour apparently motivated neither by kin-selection nor by 'reciprocal altruism'. Humanity has evolved with rational capacities because of their survival value, but rationality requires like cases to be judged alike, and to be treated alike where no relevant differences can be found. Hence the sympathies appropriate to family and friends become enlarged as reason discloses similar interests, needs or vulnerabilities elsewhere. Reason has thus expanded the circle of creatures capable of being taken into consideration so

---

7 See Attfield, *Environmental Ethics*, pp. 144–9; also Tim O'Riordan and James Cameron (eds) (1994), *Interpreting the Precautionary Principle*.

8 Willem B. Drees (2002), 'Disorder and the Ambitions of 'Science and Theology': Ten Theses', in Niels Henrik Gregerson and Ulf Görman, *Design and Disorder*, 203–23, p. 220.

9 Richard Dawkins (1976), *The Selfish Gene*, p. 205.

as to match the range of beings capable of being benefited or harmed by human activity.[10] If Singer is right, then the scope of human responsibilities may after all, as Jonas argues, correspond to that of human powers. Indeed if (as was suggested in Chapter Eight above) moral considerability belongs to non-sentient creatures and their flourishing has intrinsic value (a view that Singer rejects),[11] then his theory makes possible corresponding action grounded in these biocentric considerations alone, as well as for the sake of the human and animal interests that Singer recognizes as morally relevant. Thus defenders of stewardship should not be troubled by the objection from sociobiology under discussion here.

Future generations, however, figure not only as beneficiaries of stewardship but also as potential partners of the current generation, which can also be regarded as a partner in stewardship of previous generations. For coming generations (just like the current generation) can conserve or preserve valuable creatures and systems for their own successors, continuing the work of current stewards, just as current stewards can continue that of past ones, or can destroy such creatures and systems, whether actively or through neglect. In some cases, such as with curtailing concentrations of greenhouse gases in the atmosphere, coming generations can continue and strengthen policies of mitigation, or undermine the benefits that would be generated if current efforts were extended into the future. This means that to some extent the true subject of stewardship is an intergenerational collective, in which the later members have yet to agree to participate.

Sceptics might at this point abandon the project because of the uncertainty that future people will persevere sufficiently; but while such despair would predictably discourage coming generations from participation, efforts to transmit a sense of the values at stake cannot reasonably be denied some prospect of success. Indeed such will probably be the content of parallel debates among successive generations. Fortunately this is far from the only case where great goods can be attained if agents act in the confident belief that others in different situations will take parallel action, and recognize responsibilities to do so.[12] But the extent of time over which different agents would need to act, plus its unavoidable dependence on global inter-cultural as well as intergenerational collaboration, makes stewardship of the planet and its environs something of a special case, as well as a pressing one.

This raises the question to whom (if to anyone) the responsibilities of stewards are owed, and thus to whom stewards themselves are answerable. It can be argued that not all obligations or responsibilities are owed to anyone at all; for it can be obligatory to do something simply because it would be wrong not to do it. While I believe this view is correct, and for the reason given, the issue of answerability is not extinguished thereby, but warrants further reflection. But this brings us to the metaphysics of stewardship.

---

10 Peter Singer (1981), *The Expanding Circle*.

11 See Peter Singer [1979] (1993), *Practical Ethics* (2nd edn), 264–88.

12 See Derek Parfit (1984), *Reasons and Persons*, pp. 67–86.

Section 3: On the Metaphysics of Stewardship

Stewardship presupposes both answerability and a distinctive role for humanity. While it might seem that secular versions of stewardship discard answerability, they need not do so; as was remarked in Chapter Nine above, answerability could be owed to humanity, or (since not quite all humans are moral agents, but the moral agents of all generations are eligible to share in stewardship) to the intergenerational moral community. Being able to give an account of our stewardship to current members of this intergenerational community would be a reasonable requirement, since this is the community carrying the overall burden of responsibility for the care of the planet on its shoulders.

More than one metaphysical issue now arises. The distinctive role of humanity credited in stewardship beliefs has been held to represent a pre-evolutionary understanding of humanity and of nature, with humanity set apart in a manner irreconcilable with Darwinism.[13] Further, the implications about answerability of theistic and secular versions of stewardship raise issues concerning the possible redundancy of the former and the adequacy or otherwise of the latter, and thus of their relations and their relative credibility.

Criticisms of the presuppositions of stewardship regarding its understanding of humanity and of nature would carry weight if belief in stewardship involved either regarding humanity as independent of the evolutionary process, or alternatively as somehow indispensable at all times and places for its management. For Darwinism in all its versions involves the full participation of humanity in evolution, and also the occurrence of most of evolutionary history before humanity came on the scene. But adherents of stewardship can gladly accept these implications of Darwinism. They can accept that the powers of humans are continuous with those of their evolutionary ancestors, even though they now confer distinctive capacities and responsibilities, and that the scope of these responsibilities corresponds at any given time to the sphere of human influence and impacts, which is now extensive, though it was once slender, and was non-existent for most of the planetary past. Such adherents characteristically reject the view that everything exists for the sake of humanity, and the related belief that nature is valueless until modified by human beings. If stewardship involved universal management, these criticisms would begin to stand up; in fact, it involves the kinds of conservation and preservation that respect natural creatures and natural processes. According to theistic versions, this is the work entrusted to us by God, and in performing it our lives attain meaning of the kind derived from willingly and obediently furthering the creator's purposes.[14]

However, secular versions of stewardship can also treat the care of the planet as entrusted to the present generation of moral agents by the human community

---

[13] C. Palmer (1992), 'Stewardship: A Case Study in Environmental Ethics', in J. Ball, M. Goodhall, C. Palmer and J. Reader (eds), *The Earth Beneath*, 67-86, pp. 78–9.

[14] Drees, op. cit., p. 220.

(past, present and future). This approach well conveys that current action has moral limits, including future needs, as well as the needs of other species. While there are practical difficulties in identifying agencies to which current individuals, corporations and countries owe an account of their activities (until the United Nations or some other international body is better geared to global governance than at present), it cannot be claimed that there is no counterpart in secular stewardship to the answerability that more obviously characterizes theistic versions. (Indeed there is nothing to prevent theistic versions recognizing answerability both to God and to humanity.) It could even be suggested that secular stewardship copes so well with the problem of answerability as to make religious versions of stewardship redundant.

Theistic versions, however, incorporate metaphysical claims liable to reinforce both the responsibility and the answerability that belong to stewardship.[15] For they characteristically claim that the Earth (and analogously the rest of the universe) belong(s) to God (as in Psalm 24: 1: 'The Earth is the Lord's'), and that the land is not owned by humanity but held as a leasehold, and subject to ethical constraints (as in Leviticus 25 and Deuteronomy 15). (Thus the Old Testament passages about human dominion (Genesis 1, Psalm 8) have to be interpreted as involving just such a conditional basis.) These claims about ownership (or rather about its absence) have an ethical thrust among their practical implications, implying that people are not free to treat nature as they please. But they cannot be regarded as entirely metaphorical, or as merely ethics in disguise. Islamic teaching is to be interpreted along parallel lines.[16] Thus for theistic believers, grounds exist for the responsibilities of stewardship, to which there is no counterpart in secular versions of stewardship. Certainly Karl Marx wrote that the current generation of humanity does not own the Earth, but that as its current users they must hand it down to succeeding generations in an improved condition;[17] and this qualifies him as an adherent of a secular version of stewardship, albeit of an anthropocentric kind. But without a positive account of both answerability and ownership, this version remains less likely to motivate behaviour consonant with stewardship than some theistic versions.

Perhaps no fully satisfactory account of answerability or of ownership is to be had, and the ethics of stewardship has to motivate compliance on the basis of stewardship itself, or on that of the values that are at stake (see Chapters Eight and Nine above). Certainly the fact of widespread adherence to stewardship beliefs cannot serve as a ground for the kind of theistic belief that might in turn be capable of inspiring adherence to such beliefs. However, most theistic versions of stewardship remain more likely to be self-sustaining than most secular versions, simply because of this motivational component.

---

15 Thus Rowan Williams (2002), Dimbleby Lecture, BBC2 TV, 19 December.

16 Thus Al-Hafiz B.A. Masri (1992), 'Islam and Ecology', in Fazlun Khalid and Joanne O'Brien (eds), *Islam and Ecology*, 1–23.

17 Karl Marx [1867–94] (1967), *Capital*, vol. 3, p. 776.

Further, since there are (as argued above in Chapters Five to Eight) positive grounds for theistic beliefs that are not undermined by standard objections (including those grounded in evolution), theistic versions of stewardship should not be regarded as mere wishful thinking. (A realist interpretation of theistic beliefs is clearly crucial here, since on any other interpretation no one could be answerable to God; but, as was seen in Chapter Two, there is every reason to accept realism in the field of theistic religious language as well as in other fields.) Some aspects of theistic versions that supplement basic belief in God (including the creator's love of value, the purposefulness of evolution, and the creator's endowment of creatures including humanity with creative capacities) were found capable of rational support above in Chapter Eight; and the additional claim that the creator entrusts to his or her human creatures the care of the segments of the natural world and of its value that they can influence involves no drastic leap in the dark, but coheres so well with the previous ones as to comprise a reasonable inference if they are accepted.

Here, perhaps, we can at last discern an expression of the aspect of belief in creation mentioned in Flew and MacKinnon's dialogue[18] (see Chapter One, Section 6 above). The sense of creatureliness recognized there may not form an essential aspect of belief in creation; yet it permeates theistic versions of stewardship beliefs, conveying that because the Earth, its creatures and its environs are parts of God's creation, they belong to God rather than to humanity, and that there are related ethical responsibilities and constraints where the human treatment of it are concerned. If the creator loves value as suggested in Chapter Eight, then this will be relevant to the content of these responsibilities and constraints. Their content can be derived and discovered from reflection on values, independently of theology; but once again theistic versions turn out to have a broader coherence, and a greater capacity to inspire the kinds of commitment that are likely to be needed. What is more, as was mentioned both in Chapter One and earlier in the current section, for many they supply life with its meaning.

Needless to say, what is said here of theistic versions holds alike for Judaism, Christianity and Islam; for the passages referred to above are acknowledged in all these theistic religions; no claims to distinctive revelations have been appealed to; and the general grounds offered in earlier chapters for theistic belief support theism generically, rather than any of its specific versions in particular. The same applies to the discussion and the reasoning of the current section.

Section 4: Givenness, Metaphysics and Stewardship

A characteristic attitude of adherents of stewardship is a sense of thankfulness or gratitude. This attitude focuses partly on the privilege of caring for valuable

---

[18] Antony Flew and D.M. MacKinnon (1955), 'Creation', in Flew and McIntyre, *New Essays*, p. 173.

creatures with a view to passing on their lineages and habitats to coming generations, and partly on the sheer givenness of these entities themselves, which are thus seen as gifts.

With theistic holders of such beliefs, these feelings can readily be understood as gratitude to God the creator for conferring these valuable gifts, and for granting the privileges of life[19] and of the opportunity to care for such a trust. That the created order is not owned by humanity does not detract from awareness that its bestowal amounts to a gift, even if humanity should be considered more as a tenant than as an owner. This awareness is often combined with wonder at the intricacy, variety and magnificence of the creator's handiwork.

With adherents of secular versions of stewardship, however, these attitudes (which remain widespread) are harder to understand. The sense of privilege is not premised on anyone having granted the privilege, and the sense of thankfulness is targeted either towards abstractions such as fortune, or personifications such as Dawkins's 'blind watchmaker' or Dennett's pantheistic 'Tree of Life',[20] or towards nothing in particular. The intricacy, variety and magnificence of the world of nature can be affirmed, as Dennett remarks, but no source that can be praised for this is envisaged.[21] (Dennett's attitudes can be mentioned here because his chapter 'In Praise of Biodiversity' almost certainly qualifies him as an adherent of a secular version of stewardship.) Readers may be reminded of Ruse's 'argument to adaptive complexity' in face of nature's amazingly complex biological interactions[22] (see Chapter Eight, Section 3 above), except that Ruse was prepared to take seriously a related theology of nature.

Attitudes of thankfulness for biodiversity cohere well with, and exemplify, the biophilia hypothesis of Edward O. Wilson, who defines 'biophilia' as 'the innately emotional affiliation of human beings towards other living organisms'.[23] Wilson attempts to derive what he believes to be the fact of biophilia from his sociobiological theory of selfish genes. But it is difficult to reconcile these two theories, for, as Holmes Rolston argues, even if selfish genes could explain anything beyond their own survival, they could hardly promote human welfare of the enriched, biophiliac variety depicted in Wilson's theory, let alone biological 'diversity not their own'; for, according to the theory, they do and must produce behaviour conducive to their own survival and reproduction at the expense of the survival and reproduction of everything else. Indeed there is no need for those impressed by Wilson's claims about biophilia to endorse his genetic theories

---

19 John Cottingham (2003), *On the Meaning of Life*, p. 90.

20 See Daniel Dennett (1995), *Darwin's Dangerous Idea*, p. 520.

21 Dennett, ibid., p. 520.

22 Michael Ruse (2003), *Darwin and Design*, pp. 160-62.

23 Edward O. Wilson (1993), 'Biophilia and the Conservation Ethic', in Stephen R. Kellert and Edward O. Wilson (eds), *The Biophilia Hypothesis*, 31–41, p. 31.

of human psychology.[24] While there is some doubt about the universal extent and unqualified strength of biophilia,[25] Wilson has undoubtedly hit upon a widespread phenomenon, sufficient at least to raise the question (not resolvable here) of whether human beings in general inherit a need to associate with living creatures, and are impoverished if deprived of this association. Such an association can certainly generate or enhance a sense of life's meaningfulness,[26] particularly when combined with a sense of priorities and a concern for stewardship and conservation, however conceptualized.

However, even when such love of natural creatures and wonder at their complexity and variety takes the form of thankfulness, it would be a fallacy to argue that the occurrence of thankfulness shows there must be someone to thank. Historically, no doubt, secular stewardship derives from theistic traditions in which (as in the poetry and meditations of Thomas Traherne) to 'enjoy the world aright' is to see it as expressing God's glory;[27] and moving towards such perspectives remains open to adherents of (hitherto) secular stewardship whose sense of thankfulness is perceived to lack a focus. But secular stewardship, with its view of the care of the precious natural world as a trust owed to humanity, is not best understood as a kind of theistic stewardship in waiting, not yet understood as such by its own adherents. For this view can be allied to a secular awareness of givenness, for which the giver is nothing but the evolutionary process, and its creatures' design is undesigned.

Nor can the greater apparent scope and coherence of theistic stewardship beliefs, with their related sense of gratitude to the creator, be treated of themselves as an argument to that creator's existence. Indeed theistic stewardship beliefs (as was mentioned in the previous section) bestride several major religions, none of which need be accepted in full for these beliefs to be recognized as coherent and cogent. The generic theistic component of these beliefs, I have argued, is rationally supportable, but that support turns on its greater general explanatory adequacy (see Chapters Five to Eight above), and not on theistic stewardship beliefs having a readier account to offer of the undoubted phenomenon of cosmic gratitude. But for those who endorse belief in creation, whether on the basis of rational grounds or not, this gratitude can be recognized as due to the creator, whose gifts and generation of value form the framework that gives life its meaning.

---

[24] Holmes Rolston (1993), 'Biophilia, Selfish Genes, Shared Values', in Kellert and Wilson, 381–414, pp. 412–13.

[25] Michael E. Soulé (1993), 'Biophilia: Unanswered Questions', in Kellert and Wilson, 441–45.

[26] Ruse, *Darwin and Design*, p. 335; Kathleen Dean Moore (2005), 'The Truth of the Barnacles: Rachel Carson and the Moral Significance of Wonder', *Environmental Ethics*, **27**.3, 265–77.

[27] Thomas Traherne (1966), *Poems, Centuries and Three Thanksgivings*, 'The First Century', 28–31, 38, pp. 177–8, 181; see also 'The Salutation', pp. 5–6, and 'Wonder', pp. 6–8.

Section 5: Creation, Evolution and Meaning

The first part of this book concerned meaning and creation. The claims of critics of religious language were found to be so far from showing it to be meaningless that the central advocate of this claim, A.J. Ayer, could not avoid employing such language himself when depicting his own near-to-death experience. Both the positivist verificationism of Ayer and the pragmatist verificationism of C.S. Peirce (himself a theist) were found to be open in any case to serious objections. This finding later proved useful in contesting positivist scepticism about reasons for action, which surfaced in Chapter Eight. The meaning of belief in creation was then introduced, together with the need to understand much language used of God in a carefully qualified, analogous sense, rather than either in an over-literal, anthropomorphic sense or in an unspecified and equivocal sense; otherwise objections to theism that contrast the world's evils and God's goodness, later considered in Chapters Six and Seven, would not make sense in the first place. Attitudes to creation often treated as integral to theistic belief were remarked and saved for later consideration when issues of stewardship came under discussion in Chapter Ten.

Currently a pressing issue in philosophy of religion (as in several branches of philosophy) is that of realism and anti-realism, which was found in the second chapter to be fundamentally a debate about meaning. The principal arguments of Michael Dummett and Crispin Wright for generic anti-realism were presented but shown to be inconclusive, as was the truth-conditional theory of meaning (sometimes used in support of anti-realism); and so too was the variety of anti-realism held by Richard Rorty, who claims that the world is created by language-users, rather than by God. This kind of anti-realism is applied both by Rorty and by Don Cupitt to creation and the creator, but is found to supply an unsatisfactory account of ordinary perceptual and scientific discovery (for example, of the North Downs), let alone of religious language. The distinctive religious anti-realism of D.Z. Phillips was also found wanting; while belief in creation can give meaning to the life of the believer, it can hardly do so if no creator and no creative purposes figure in this belief. The arguments of Peter Byrne were endorsed for accepting a realist theory in matters of understanding theistic religious language, including the language of creation; and a critical realist stance emerged as appropriate for the philosophy of science as well as the philosophy of religion. A realist stance turns out to be vital for rational belief in God as creator (see Chapters Four to Seven) and as source of the world's value (see Chapter Eight), and for human answerability to God (as well as to humanity) as stewards of planetary nature (see Chapters Nine and Ten); for if God and creation were functions of human belief, and God were dependent on people rather than vice versa, then rational belief in creation would be out of the question.

In the third chapter, falsification was brought to bear on theistic belief, and shown to throw light on its logic. It was shown how falsificationists can accept Quinean holism without endorsing scepticism about meaning. Discussion of the debate surrounding Antony Flew's parable of the gardener showed that Flew tacitly admitted that religious language can be meaningful as expressing statements, while stressing that those who make them have to confront issues of theodicy (tackled here in Chapters Six and Seven). The issue of how in theory such statements could be falsified was then addressed, in a way that throws light on their meaning. Falsification, it emerged, need not be empirical, and applies to clusters of statements, rather than to statements considered singly. The bearing of Flew's more recent beliefs about God on the falsifiability of beliefs about creation was also considered; if some form of deism or Aristotelian theism[28] can serve as the best explanation of the world and of life on earth, then at least these versions of theology must be deemed falsifiable in principle. Beliefs about creation too were argued to be in principle falsifiable, and capable of being regarded as the best explanation of the physical world being as it is. This makes it worthwhile to clarify creation beliefs (the role undertaken in Chapter Four) and to supply grounds for them (the role of Chapter Five).

The second part of the book concerned creation and evolution (Chapters Four to Eight). Chapter Four revisited theistic belief in creation, contrasting it with alternatives such as Platonism and pantheism, and bringing out its implications for such topics as transcendence, contingency and time; according to this belief, the creator eternally (or timelessly) brings it about that creatures exist and that natural processes such as evolution unfold at particular times. Belief in creation was then contrasted with a belief with which it is sometimes confused, creationism, of which both nineteenth-century and contemporary forms were discussed. Finally the possibility of believing in creation while rejecting creationism was illustrated from the beliefs of the middle decades of the life of Charles Darwin. This possibility has been further explored in the rest of the book.

Chapter Five presented and discussed grounds for belief in God and in creation. The first two sections concerned the Cosmological Argument, which concludes that all material objects are, have been or will be dependent on a creative agent with the power to make the difference between their existence and their non-existence, or to bring them about. The world is not argued to have had a temporal origin, but to be dependent on God at all times. The other three sections discussed arguments from design, rejecting (on Humean grounds) arguments from biological adaptations, however complex (such as the eye), but defending arguments from 'regularities of succession' such as laws of nature (of which evolution by natural selection turns out to be one), and also the Argument from Fine Tuning. Hume's objections, powerful as they are to purposive explanations of particular phenomena, were found inconclusive with regard to such explanations at the level of laws and constants, and thus of the framework within which life has evolved. Here, Darwinism positively strengthens the design argument from regularities of succession by itself supplying a further example of such a regularity.

---

[28] Both these phrases are used by Flew of his current beliefs about God: see Antony Flew (2005), 'My "Conversion"', *Think*, **11**, pp. 79, 80.

Further, the Argument from Fine Tuning turned out to remain persuasive despite recent criticisms of the concept of probability that it employs. Consideration of Hume's objections to belief in God's goodness was deferred until after a more detailed discussion of Darwinism in Chapter Six; yet a strong case for belief in God and creation was already apparent.

Chapter Six opened with a review of the varieties of Darwinism and the grounds for accepting its central claims. Varieties introduced here like those of Stephen Jay Gould and of Richard Lewontin that do not wholly turn on adaptations become important later, alongside the all-encompassing adaptationism of Richard Dawkins. The following two sections considered the compatibility of Darwinism with belief in creation, and concluded (along lines parallel to the recent discussions on the parts of Gould and of Michael Ruse) that there is no clear incompatibility. However, the objections to belief in God's goodness deriving from Hume and from Darwinism now required to be tackled. The next section sifted Humean objections, some of them echoed by the later Darwin, but these objections proved inconclusive, while the final section discussed the claims of some Darwinists about nature's immorality, together with the apparent problems of predation, waste and parasitism. The suggestion that a world without these phenomena would be better was challenged here, in view of the systems for which what is sometimes considered waste is vital, and the kinds of life that could not exist in the absence of natural selection with its predation and parasitism. Here a systemic view of evolution proves indispensable for purposes of theodicy.

In Chapter Seven the relation of belief in God to recognition of the fact of evil was addressed more fully. The opening section argued that these beliefs involve no inconsistency, and that moral evil could be reconciled with belief in God's goodness along the lines of the Free Will Defence. In the following section, a world of miraculous interventions to countermand atrocities was argued to be less desirable than a law-governed world; if so, a purposive creator, desirous of value, would be unlikely to intervene to avert humanly originated catastrophes. The three remaining sections considered the problem of natural evil. The first of them remarked that the issue turns on whether alternative possible natural regularities would be preferable; before it can be asserted that they would be, the far-flung implications across the whole of space and time, both welcome and unwelcome, would have to be taken into account. The next discussed the problem as posed by Richard Dawkins, and the adequacy of an ingenious reply from Michael Ruse: maybe the kinds of life that we value could not have emerged without natural selection (and all the predation and parasitism that this implies), and so the creator had no other resort. The final section considered theories supplementary to that of Dawkins, and also theories for which evolution is not driven by adaptation only, and concluded that all these possible theories both invoke natural selection and are compatible with belief in divine creation. Here again, reflection on evolution by natural selection provides vital steps in an argument for the compatibility of creation and observable phenomena.

Chapter Eight presented and discussed Keith Ward's argument from value, and his 'theistic evolution' stance. His account of intrinsic value turned out to stand in need of being radically reconceptualized, and the argument from value of being revised accordingly; this was done in the two opening sections, where a distinctive theory of value was introduced. In the next two sections, the revised argument was held to supply independent support for belief both in creation by a creator desirous of a world of intrinsic value and in evolution being progressive in accordance with the purposes of such a creator. While objections to such a progressivist argument from philosophers such as Ruse were considered, diagnosed and rejected, a parallel argument to that from value was found in Ruse's own Argument from Adaptive Complexity, according to which the wonder generated by observation of adaptive complexity in all its variety reasonably generates a religious response, consistent with belief in creation. Here once again, Darwinism turns out to contribute to aspects of belief in creation. The final section argued that this account of creation is in turn supportive of God's immanence in the created order and of a panentheist view of God, a view that became much more cogent when, as Darwinism became accepted, naturalistic processes were recognized to contribute to the work of creation.

The third part of the book concerned evolution and meaning. Chapter Nine attempted to relate linguistic meaning to understandings of the meaning of life, developing themes of the opening section of Chapter One. While meaning is not to be equated with value, there turned out to be important links between these concepts; some senses of 'meaning', however, apply only to agents with a concept of themselves. The following two sections discussed the possibility that meaning has itself evolved alongside the evolutionary development of complexity, along the lines suggested by Daniel Dennett. His engaging account, however, stands in need of correctives, and the sense in which meaning could have evolved turns out neither to be deterministic nor to involve any straightforward extension of Darwinism. The following sections concerned what is required for life to be meaningful (including integrated priorities, self-awareness and a sense of objective values), and then the possibilities for their satisfaction by adherents of different varieties of Darwinism (or Darwinian metaphysic). The final section of Chapter Nine discussed and defended one set of beliefs and attitudes that can make life meaningful, belief in the human stewardship of the natural world of the planet and its environs. This involves care of the fruits of evolution, and can be found both in religious and secular versions; Darwinism can nourish either of these versions through the wonder evoked by contemplation of these fruits.

In Chapter Ten, ethical and metaphysical aspects of stewardship were considered. Ethical aspects include grounds for an ethic of stewardship and the possibility of compliance with it in view of our evolutionary inheritance. Metaphysical aspects include the understandings of nature and of humanity implicit in stewardship beliefs, and issues surrounding the related beliefs in answerability. Theistic versions were found to have greater coherence, but adopting them on a rational basis would involve having regard to the arguments for and against theistic belief (discussed in Chapters Five to Eight), rather than on the strength of this coherence alone. Much the same applies to the widespread phenomenon of thankfulness for life and the value of living creatures, discussed in the following section, and to accepting belief in a creator to whom such thankfulness could be directed. This widespread aspect of stewardship belief (thankfulness for life and its value) exemplifies the condition of biophilia which the sociobiologist E.O. Wilson has claimed to pervade humanity, and which he grounds in our evolutionary history. The phenomenon of biophilia would be hard to derive, it was argued, from his sociobiology and its genetic reductionism. It would cohere rather with attitudes of wonder at nature's variety, complexity and magnificence, attitudes fostered, as we have seen, by acceptance of Darwinism, and which readily contribute to a sense of life's meaningfulness when combined with belief in stewardship, whether theistic or secular.

The meaning of life, as John Cottingham claims, is a hermeneutic concept,[29] depending for its content on how life is interpreted. But this cannot be a purely subjective matter, as issues of what is objectively valuable are involved (see Chapters Eight and Nine). While some beliefs, belief in evolution among them, have sometimes been regarded as robbing life of its meaning, none of the varieties of Darwinian metaphysic need be so interpreted (as was argued in Chapter Nine), neither Mind-First Darwinism, nor moderate materialist Darwinism, nor Gene-Machine Darwinism, even though some interpretations of the latter suggest otherwise. Other beliefs again, including belief in creation (Chapter Four) and in the creator's good purposes (Chapter Eight), supply a distinctive frame of reference contributory to life being discerned as meaningful (Chapter Ten), although their adoption simply for the sake of achieving a sense of life's meaningfulness could well prove counter-productive. These beliefs have here been argued to survive charges of meaninglessness (Chapter One), to be amenable to realist interpretation (Chapter Two), and to be compatible with the central themes of Darwinism (Chapters Six and Seven). The objections to these beliefs turn out to be inconclusive, and good grounds for holding them turn out to exist. No rational barrier (at least as far as the issues discussed above are concerned) should therefore be taken to debar their adoption. Far from being a barrier, a commitment to Darwinism illuminates, facilitates and can serve to strengthen these beliefs.

---

29 John Cottingham (2003), *On the Meaning of Life*, pp. 21–2.

Bibliography

Alston, William (2003), 'Religious Language and Verificationism', in Copan and Moser, *The Rationality of Theism* (see below), 17–34

Atkins, Peter (1981), *The Creation*, Oxford and San Francisco: W.H. Freeman

Attfield, Robin (1970), 'Non-Tentative Religious Beliefs and Rationality', *Sophia*, **9**.1, 16–21

Attfield, Robin (1971), 'The Individuality of God', *Sophia*, **10**.1, 20–27

Attfield, Robin (1974), 'On Being Human', *Inquiry*, **17**, 175–92

Attfield, Robin (1980), 'Religious Symbols and the Voyage of Analogy', *International Journal of the Philosophy of Religion*, **11**, 225–38

Attfield, Robin (1983), 'Necessity and Contingency in God', *New Blackfriars*, **65**, 35–41

Attfield, Robin (1984), 'Work and the Human Essence', *Journal of Applied Philosophy*, **1**.1, 141–50

Attfield, Robin [1978] (1993), *God and the Secular: A Philosophical Assessment of Secular Reasoning from Bacon to Kant* (2nd edn), Aldershot and Brookfield, VT: Gregg Revivals

Attfield, Robin (1993), 'Clarke, Independence and Necessity', *British Journal for the History of Philosophy*, **1**.2, 67–82

Attfield, Robin (1994), *Environmental Philosophy: Principles and Prospects*, Aldershot and Brookfield, VT: Ashgate

Attfield, Robin (1995), *Value, Obligation and Meta-Ethics*, Amsterdam and Atlanta, GA: Éditions Rodopi

Attfield, Robin (1996), 'Too High a Theme? Of Finitude, Predication and Analogy', *Scottish Journal of Religious Studies*, **17**, 5–19

Attfield, Robin (1999), *The Ethics of the Global Environment*, Edinburgh: Edinburgh University Press, and West Lafayette, IN: Purdue University Press

Attfield, Robin (2000), 'Evolution, Theodicy and Value', *The Heythrop Journal*, **41**.3, 281–96

Attfield, Robin (2001), 'Postmodernism, Value and Objectivity', *Environmental Values*, **10**.2, 145–62

Attfield, Robin (2001), 'Meaningful Work and Full Employment', *Reason in Practice*, **1**.1, 41–48

Attfield, Robin (2003), *Environmental Ethics: An Overview for the Twenty-first Century*, Cambridge: Polity and Malden, MA: Blackwell

Attfield, Robin (2004), 'Rousseau, Clarke, Butler and Deism', *British Journal for the History of Philosophy*, **12**.3, 429–43

Austin, J.L. (1970), *Philosophical Papers* (2nd edn), ed. J.O. Urmson and G.J. Warnock, London: Oxford University Press

Ayer, A.J. [1936] (1946), *Language, Truth and Logic* (2nd edn), London: Victor Gollancz

Ayer, A.J. (1973), *The Central Questions of Philosophy*, London: Weidenfeld and Nicolson

Ayer, A.J. (1988), 'What I Saw When I Was Dead', *Sunday Telegraph*, 28 August

Ayer, A.J. (1988), 'Postscript to a Postmortem', *Spectator*, 15 October

Ayer, A.J. (1992), 'Still More of My Life' in Hahn, *The Philosophy of A.J. Ayer* (see below), 44–50

Ayer, A.J. (1992), 'Reply to Michael Dummett', in Hahn, *The Philosophy of A.J. Ayer* (see below), 149–56

Ball, J., M. Goodhall, C. Palmer and J. Reader (eds) (1992), *The Earth Beneath*, London: SPCK

Banner, Michael C. (1990), *The Justification of Science and the Rationality of Religious Belief*, Oxford: Clarendon Press

Barbour, Ian G. (1998), *Religion and Science: Historical and Contemporary Issues*, London: SCM Press

Barrow, John D. and Frank J. Tipler (1986), *The Anthropic Cosmological Principle*, Oxford: Oxford University Press

Bartholomew, D.J. (1984), *God of Chance*, London: SCM

Beer, Gillian (1983), *Darwin's Plots: Evolutionary Narrative in Darwin, George Eliot, and Nineteenth-century Fiction*, London; Boston: Routledge and Kegan Paul

Behe, Michael (1996), *Darwin's Black Box: The Biochemical Challenge to Evolution*, New York: Free Press

Behe, Michael (2004), 'Irreducible Complexity: Obstacle to Darwinian Evolution', in Dembski and Ruse, *Debating Design* (see below), 352–70

Bendall, D.S. (ed.) (1983), *Evolution from Molecules to Men*, Cambridge: Cambridge University Press

Berkeley, George [1732] (1950), *Alciphron, or the Minute Philosopher*, in George Berkeley (1948–57), *Works*, ed. A.A. Luce and T. Jessop, London: Thomas Nelson and Sons, vol. III

Berry, R.J. (1988), *God and Evolution*, London: Hodder and Stoughton

Bhaskar, Roy (1986), *Scientific Realism and Human Emancipation*, London: Verso

Blackstone, William T. (ed.) (1974), *Philosophy and Environmental Crisis*, Athens, GA: University of Georgia Press

Boethius (1897), *De Consolatione Philosophiae*, trans. George Colville, ed. Ernest Belfort Bax, London: Nutt

Boghossian, Paul (1989), 'The Rule-following Considerations', *Mind*, **98**, 507–49

Braithwaite, R.B. (1955), *An Empiricist's View of the Nature of Religious Belief*, The Ninth Arthur Stanley Eddington Memorial Lecture, Cambridge: Cambridge University Press

Brooke, John Hedley (1985), 'The Relations Between Darwin's Science and His Religion', in Durant, *Darwinism and Divinity* (see below), 40–75

Brooke, John Hedley (2002), 'Revisiting Darwin on Order and Design', in Gregerson and Görman, *Design and Disorder* (see below), 31–52

Brown, Andrew (1999), *The Darwin Wars: How Stupid Genes Became Selfish Gods*, London: Simon and Schuster

Brown, Patterson (1964), 'St. Thomas' Doctrine of Necessary Being', *Philosophical Review*, **73**, 76–90

Brown, Peter G. (2000), *Ethics, Economics and International Relations: Transparent Sovereignty in the Commonwealth of Life*, Edinburgh: Edinburgh University Press

Burrell, David (1973), *Analogy and Philosophical Language*, New Haven, CT and London: Yale University Press

Byrne, Peter (2003), *God and Realism*, Aldershot and Burlington, VT: Ashgate

Carnap, Rudolph [1928] (1967), *The Logical Structure of the World*, trans. R.A. George, London: Routledge and Kegan Paul

Carrier, Richard (2005), 'Antony Flew Considers God … Sort Of', <http://www.secweb.org/asset.asp?AssetID=369>, most recently updated January 2005, when visited in October 2005

Chadwick, Owen (1966), 'Evolution and the Churches', in Owen Chadwick, *The Victorian Church*, Part II, London: A. and C. Black, 23–35

Chadwick, Owen [1966] (1973), 'Evolution and the Churches', in C.A. Russell, *Science and Religious Belief* (see below), 282–93

Clark, Stephen R.L. (2000), 'The Evolution of Language: Truth and Lies', *Philosophy*, **75**, 401–21

Clayton, Philip (1997), *God and Contemporary Science*, Edinburgh: Edinburgh University Press

Clayton, Philip (2000), *The Problem of God in Modern Thought*, Grand Rapids, MI: William B. Eerdmans

Clayton, Philip and Arthur Peacocke (eds) (2004), *In Whom We Live and Move and Have Our Being*, Grand Rapids, MI and Cambridge: William B. Eerdmans

Conway, David (2000), *The Rediscovery of Wisdom: From Here to Antiquity in Quest of Sophia*, Basingstoke and New York: Palgrave

Cooper, David E. (2003), *Meaning*, Chesham: Acumen

Copan, Paul and Paul K. Moser (eds) (2003), *The Rationality of Theism*, London and New York: Routledge

Copan, Paul and William Lane Craig (2004), *Creation out of Nothing: A Biblical, Philosophical and Scientific Exploration*, Grand Rapids, MI: Baker Academic

Cottingham, John (2003), *On the Meaning of Life*, London and New York: Routledge

Craig, William Lane (1993), 'A Criticism of the Cosmological Argument for God's Non-Existence', in Craig and Smith, *Theism, Atheism and Big Bang Cosmology* (see below), 256–76

Craig, Willliam Lane and Quentin Smith (1993), *Theism, Atheism and Big Bang Cosmology*, Oxford: Clarendon Press

Craig, William Lane (1998), 'Divine Timelessness and Personhood', *International Journal for the Philosophy of Religion*, 43, 109–24

Cross, F.L. and E.A. Livingstone (eds) (1974), *The Oxford Dictionary of the Christian Church*, Oxford: Oxford University Press

Cupitt, Don (1986), *Life Lines*, London: SCM Press

Cupitt, Don (1990), *Creation out of Nothing*, London: SCM

Cupitt, Don (1993), 'Anti-Realist Faith', in Runzo, *Is God Real?* (see below), 45–55

Darwin, Charles (1872), *The Origin of Species* (Darwin's last edition), London: John Murray

Darwin, Francis (ed.) (1888), *The Life and Letters of Charles Darwin* (3 vols), London: John Murray

Davies, Brian (1982), 'Kenny on God', *Philosophy*, **57**, 105–17

Davies, Brian (1983), 'A Timeless God', *New Blackfriars*, **64**, 215–24

Davis, Stephen T. (1981), 'Free Will and Evil', in Davis, *Encountering Evil* (see below), 69–83

Davis, Stephen T. (1981) (ed.), *Encountering Evil: Live Options in Theodicy*, Edinburgh: T. and T. Clark

Davis, Stephen T. (1993), 'Against "Anti-Religious Faith"', in Runzo, *Is God Real?* (see below), 57–59

Davis, Stephen T. (1997) *God, Reason and Theistic Proofs*, Edinburgh: Edinburgh University Press

Dawkins, Richard (1976), *The Selfish Gene*, Oxford: Oxford University Press

Dawkins, Richard (1981), *The Extended Phenotype: The Gene as the Unit of Selection,* Oxford: Freeman

Dawkins, Richard (1983), 'Universal Darwinism', in Bendall, *Evolution from Molecules to Men* (see above), 403–25

Dawkins, Richard (1986), *The Blind Watchmaker*, London: Penguin

Dawkins, Richard (1989), *The Selfish Gene* (2nd edn), Oxford: Oxford University Press

Dawkins, Richard (1995), *A River Out of Eden*, New York: Basic Books

Dawkins, Richard (1996), *Climbing Mount Improbable*, London: Viking

Dawkins, Richard (2004), *A Devil's Chaplain: Selected Essays*, London: Phoenix

de Beer, Gavin (1960), 'Darwin's Notebooks on the Transmutation of Species', *Bulletin of the British Museum (Natural History)*, Historical Series, **2**, parts 2–5

de Waal, Frans (1996), *Good Natured: The Origins of Right and Wrong in Humans and Other Animals*, Cambridge, MA and London: Harvard University Press

Dembski, William A. (1998), *The Design Inference: Eliminating Chance through Small Probabilities*, Cambridge: Cambridge University Press

Dembski, William A. (2003), *The Design Revolution: Answering the Toughest Questions about Intelligent Design*, Downers Grove, IL and Leicester: InterVarsity Press

Dembski, William A. and Michael Ruse (eds) (2004), *Debating Design: From Darwin to DNA*, Cambridge: Cambridge University Press, 81–97

Dennett, Daniel (1991), *Consciousness Explained*, Boston: Little, Brown

Dennett, Daniel (1995), *Darwin's Dangerous Idea: Evolution and the Meanings of Life*, Penguin

Dennett, Daniel (2003), *Freedom Evolves*, London: Allen Lane

Devitt, Michael (1984), *Realism and Truth*, Oxford: Basil Blackwell

Dostoevsky, Fyodor (1993), *The Brothers Karamazov*, trans. David McDuff, Harmondsworth: Penguin

Drees, Willem B. (1996), *Religion, Science and Naturalism*, Cambridge: Cambridge University Press

Drees, Willem B. (2002), 'Disorder and the Ambitions of "Science and Theology": Ten Theses', in Gregerson and Görman (eds), *Design and Disorder* (see below), 203–23

Duhem, Pierre [1914] (1954), *The Aim and Structure of Physical Theory*, trans. P. Wiener, Princeton, NJ: Princeton University Press

Dummett, Michael (1973), *Frege: Philosophy of Language*, London: Duckworth

Dummett, Michael (1978), *Truth and Other Enigmas*, London: Duckworth

Dummett, Michael (1987), 'Reply to Brian Loar', in B. Taylor, *Michael Dummett: Contributions to Philosophy* (see below), 269–80

Dummett, Michael (1992), 'The Metaphysics of Verificationism', in Hahn, *The Philosophy of A.J. Ayer* (see below), 129–48

Dummett, Michael (1993), *The Seas of Language*, Oxford: Clarendon Press

Dummett, Michael [1976] (1999), 'What is a Theory of Meaning? (II)', in Evans and McDowell, *Truth and Meaning* (see below), 67–137

Dupré, John (2001), *Human Nature and the Limits of Science*, Oxford: Clarendon Press

Dupré, John (2003), *Darwin's Legacy: What Evolution Means Today*, Oxford: Oxford University Press

Durant, John (ed.) (1985), *Darwinism and Divinity: Essays on Evolution and Religious Belief*, Oxford and New York: Basil Blackwell

Durrant, Michael (1973), *The Logical Status of 'God'*, London: Macmillan

Edwards, Paul (ed.) (1977), *The Encyclopedia of Philosophy*, New York: Macmillan

Eldredge, Niles and Stephen Jay Gould (1972), 'Punctuated Equilibria: An Alternative to Phyletic Gradualism', in T.J.M. Schopf (ed.), *Models in Paleobiology*, San Francisco, CA: Freeman, Cooper, 82–115

Evanoff, Richard J. (2005), 'Reconciling Realism and Constructivism in Environmental Ethics', *Environmental Values*, **14**.1, 61–81

Evans, Gareth and John McDowell (eds) (1999), *Truth and Meaning: Essays in Semantics* (2nd edn), Oxford: Clarendon Press

Evans, J.L. (1953), 'On Meaning and Verification', *Mind*, **62**, 1–19

Farrer, A.M. (1952), 'Introduction' in G.W. Leibniz, *Theodicy*, trans. E.M. Huggard, ed. A.M. Farrer, London: Routledge and Kegan Paul

Farrer, A.M. (1964), *Saving Belief*, London: Hodder and Stoughton

Feigl, Herbert (1963), 'Physicalism, Unity of Science and the Foundations of Psychology', in P.A. Schilpp (ed.), *The Philosophy of Rudolph Carnap* (see below), 227–67

Feinberg, Joel (1974), 'The Rights of Animals and Unborn Generations', in Blackstone, *Philosophy and Environmental Crisis* (see above), 43–68

Ferré, Frederick (1961), *Language, Logic and God*, New York: Harper and Brothers

Feuerbach, Ludwig [1841] (1957), *The Essence of Christianity*, trans. M.A. Evans (George Eliot), New York: Harper and Row

Findlay, J.N. (1955), 'Can God's Existence be Disproved?', in A. Flew and A. MacIntyre (eds), *New Essays* (see below), 47–56

Flew, Antony (1955), 'Theology and Falsification, A', in Flew and MacIntyre, *New Essays* (see below), 96–99

Flew, Antony (1955), 'Theology and Falsification, D', in Flew and MacIntyre, *New Essays* (see below), 106–08

Flew, Antony and D.M. MacKinnon (1955), 'Creation', in Flew and McIntyre, *New Essays* (see below), 170–86

Flew, A. and A. McIntyre (eds) (1955), *New Essays in Philosophical Theology*, London: SCM Press

Flew, Antony (1966) *God and Philosophy*, London: Hutchinson

Flew, Antony (2004), 'On Darwinism and Theology', *Philosophy Now*, **47**, August/September, 22

Flew, Antony (2005), 'My "Conversion"', *Think: Philosophy for Everyone*, **11**, Autumn, 75–84

Foot, Philippa (1978), *Virtues and Vices*, Oxford: Blackwell

Foster, Michael (1934), 'The Christian Doctrine of Creation and the Rise of Modern Natural Science', *Mind*, **43**, 446–68

Foucault, Michel (1972), *The Archaeology of Knowledge*, trans. A.M. Sheridan Smith, New York: Harper and Row

Gallagher, Kenneth T. (1989), '"Natural Selection": A Tautology?,' *International Philosophical Quarterly*, **29**, 17–31

Geach, Peter (1969), 'Causality and Creation', in Geach, *God and the Soul* (see below), 75–85

Geach, Peter (1969), *God and the Soul*, London: Routledge and Kegan Paul

General Synod of the Church of England Board for Social Responsibility (1991), *Christians and the Environment*, London: Board for Social Responsibility

Gillespie, Neal C. (1979), *Charles Darwin and the Problem of Creation*, Chicago and London: University of Chicago Press

Gillispie, Charles Coulston (1959), *Genesis and Geology: The Impact of Scientific Discoveries upon Religious Beliefs in the Decades before Darwin*, New York: Harper and Row

Goodpaster, Kenneth E. (1978), 'On Being Morally Considerable', *Journal of Philosophy*, **75**, 308–25

Gore, Charles (ed.) (1891), *Lux Mundi*, London: Murray

Gould, Stephen Jay (1983), 'Evolution as Fact and Theory', in Stephen Jay Gould, *Hen's Teeth and Horse's Toes*, New York: Norton, 253–62

Gould, Stephen Jay (1991), *Bully for Brontosaurus*, London: Penguin

Gould, Stephen Jay and Richard C. Lewontin (1994), 'The Spandrels of San Marco and the Panglossian Paradigm: A Critique of the Adaptationist Programme', in Sober, *Conceptual Issues in Evolutionary Biology* (see below), 73–90

Gould, Stephen Jay (1997), *Life's Grandeur: The Spread of Excellence from Plato to Darwin*, Harmondsworth: Penguin

Gould, Stephen Jay and Elisabeth S. Vrba (1998), 'Exaptation: A Missing Term in the Science of Form', in Hull and Ruse, *The Philosophy of Biology* (see below), 52–71

Gould, Stephen Jay (1999), *Rocks of Ages: Science and Religion in the Fullness of Life*, New York: Ballantine

Gould, Stephen Jay (2000), 'More Things in Heaven and Earth', in Rose and Rose, *Alas, Poor Darwin* (see below), 85–105

Gregerson, Niels Henrik and Ulf Görman (eds) (2002), *Design and Disorder: Perspectives from Science and Theology*, London and New York

Grice, H.P. (1989), 'Logic and Conversation', in H.P. Grice, *Studies in the Way of Words*, Cambridge, MA and London: Harvard University Press, 22–40

Gruber, Howard E. (1974), *Darwin on Man: A Psychological Study of Scientific Creativity, Together with Darwin's Early and Unpublished Notebooks*, transcribed and annotated by Paul H. Barrett, London: Wildwood House

Hahn, L.E. (ed.) (1992), *The Philosophy of A.J. Ayer*, La Salle, IL: Open Court

Hale, Bob (1997), 'Realism and Its Oppositions', in Hale and Wright, *Companion* (see below), 271–308

Hale, Bob (1997), 'Rule-Following, Objectivity and Meaning', in Hale and Wright, *Companion* (see below), 369–96

Hale, Bob and Crispin Wright (eds) (1997), *A Companion to the Philosophy of Language*, Oxford and Malden, MA: Blackwell

Hanfling, Oswald (1987), *The Quest for Meaning*, Oxford and Cambridge, MA: Blackwell

Hannam, James (2004), Review of Stephen Barr (2003), *Modern Science and Ancient Faith*, Notre Dame, IN: Notre Dame Press, *Science and Religion Forum Reviews*, **43**, 16–18

Harding, Sandra G. (ed.) (1976), *Can Theories Be Refuted?: Essays on the Duhem-Quine Thesis*, Dordrecht and Boston, MA: D. Reidel

Hare, R.M. (1955), 'Theology and Falsification, B', in Flew and MacIntyre, *New Essays* (see above), 99–103

Hawkins, Ronnie Zoe (2005), Review of Ruse, *Darwin and Design* (see below) in *Environmental Ethics*, **27**.2, 209–12

Hebblethwaite, Brian (1993), 'A Critique of Don Cupitt's Christian Buddhism', in Runzo, *Is God Real?* (see below), 135–48

Hebblethwaite, Brian (2000) *Evil, Suffering and Religion* (revised edn), London: SPCK

Hepburn, R.W. (1963), 'From World to God', *Mind*, **72**, 40–50

Hick, John (1963), *Philosophy of Religion*, Englewood Cliffs, NJ: Prentice-Hall

Hick, John (ed.) (1964), *The Existence of God*, New York: Macmillan

Hick, John (1966), *Evil and the God of Love*, London and Basingstoke: Macmillan

Hick, John (1971), 'Theology and Verification', in Mitchell, *Philosophy of Religion* (see below), 53–71

Hick, John (1981), 'An Irenaean Theodicy', in Davis, *Encountering Evil* (see above), 39–52

Hick, John (1993), 'Believing – And Having True Beliefs', in Runzo, *Is God Real?* (see below), 115–16

Hick, John (1993) 'Belief in God: Metaphysics and Values', in Runzo, *Is God Real?* (see below), 130–32

Hill, Thomas E., Jr (1984), Review of Robert Elliot and Arran Gare (eds), *Environmental Philosophy*, University Park, PA: Pennsylvania State University Press; London: Open University Press; St Lucia: University of Queensland Press, in *Environmental Ethics*, **6**, 367–71

Hirst, R.J. (1977), 'Realism' in Edwards, *The Encyclopedia of Philosophy* (see above), 77–83

Holdcroft, David and Harry Lewis (2001), 'Consciousness, Design and Social Practice', *Journal of Consciousness Studies*, **8**.8, August, 43–58

Hook, Bill (2003), 'Intrinsic Value: Under the Scrutiny of Information and Evolutionary Theory', *Environmental Ethics*, **25**.4, 359–73

Hooykaas, R. (1972), *Religion and the Rise of Modern Science*, Edinburgh and London: Scottish Academic Press

Hooykaas, R. (1974) 'Nature and History', in Open University Course Team for Science and Belief: From Copernicus to Darwin, *The New Outlook for Science* (see below), 10–33

Horwich, Paul (1998), *Meaning*, Oxford: Clarendon Press

Huiser, Pieter J. (1997), *Models, Theories and Narratives: Conditions for the Justification of a Religious Realism*, Amsterdam: Thesis Publishers

Hull, David L. (1988), *Science as a Process*, Chicago: University of Chicago Press

Hull, David L. and Michael Ruse (eds) (1998), *The Philosophy of Biology*, Oxford: Oxford University Press

Hume, David [1779] (1963), 'Dialogues concerning Natural Religion', in Wollheim, *Hume on Religion* (see below), 99–204

Hume, David (1975), *Enquiries Concerning Human Understanding and Concerning the Principles of Morals* (3rd edn), ed. L.A. Selby-Bigge, rev. P.H. Nidditch, Oxford: Clarendon Press

Hurlbutt, Robert H. (1965), *Hume, Newton and the Design Argument*, Lincoln, NE: University of Nebraska Press

Huxley, T.H. (1894), *Evolution and Ethics and Other Essays*, New York: D. Appleton

Huxley, T.H. and Julian Huxley (1947), *Evolution and Ethics*, London: Pilot Press

Jonas, Hans (1984), *The Imperative of Responsibility: In Search of an Ethics for the Technological Age*, trans. Hans Jonas and David Herr, Chicago and London: University of Chicago Press

Kant, Immanuel [1781] (1968), *Critique of Pure Reason*, in N. Kemp Smith (trans.), *Immanuel Kant's Critique of Pure Reason*, London: Macmillan, 65–119

Kellert, Stephen R. and Edward O. Wilson (eds) (1993), *The Biophilia Hypothesis*, Washington, DC: Island Press

Kenny, Anthony (1969), *The Five Ways*, London: Routledge and Kegan Paul

Kenny, Anthony (1979), *The God of the Philosophers*, Oxford: Clarendon Press

Kenny, Anthony (2004), *The Unknown God: Agnostic Essays*, London and New York: Continuum

Kerr, Fergus (1986), *Theology after Wittgenstein*, Oxford: Blackwell

Khalid, Fazlun and Joanne O'Brien (eds) (1992), *Islam and Ecology*, London and New York: Cassell

Kingsley, Charles [1863] (1930), *The Water-Babies*, London: Hodder and Stoughton

Kitcher, Philip (1985), *Vaulting Ambition: Sociobiology and the Quest for Human Nature*, Cambridge, MA: MIT Press

Kropotkin, Petr [1902] (1955), *Mutual Aid*, ed. A. Montague, Boston, MA: Extending Horizons Books

Kuhn, Thomas S. (1962), *The Structure of Scientific Revolutions*, Chicago: University of Chicago Press

Le Poidevin, Robin (1996), *Arguing for Atheism: An Introduction to the Philosophy of Religion*, London: Routledge

Lepore, E. and B. Smith (eds) (forthcoming), *Oxford Handbook on the Philosophy of Language*, Oxford: Oxford University Press

Leslie, John (1989), *Universes*, London: Routledge

Lewis, H.D. (1959), *Our Experience of God*, London: George Allen and Unwin

Lewontin, Richard (1970), 'The Units of Evolution', *Annual Review of Ecology and Systematics*, **1**, 1–23

Lovejoy, Arthur O. (1936), *The Great Chain of Being: A Study of the History of an Idea*, Cambridge, MA: Harvard University Press

Lucas, J.R. (1970), *The Freedom of the Will*, Oxford: Clarendon Press

Lucretius, T. Carus (1947), *De Rerum Natura*, ed. and trans. Cyril Bailey (3 vols), Oxford: Clarendon Press

Lyell, Charles [1830] (1875), *Principles of Geology* (12th edn), London: John Murray

McCabe, Herbert (1992), 'The Logic of Mysticism – I', in Warner, *Religion and Philosophy* (see below), 45–59

McCormick, Peter J. (ed.) (1996), *Starmaking: Realism, Anti-Realism and Irrealism*, Cambridge, MA and London: MIT Press

McDowell, John (1992), 'Meaning and Intention in Wittgenstein's Later Philosophy', *Midwest Studies in Philosophy*, **17**, 40–52

McGrath, Alister (2005), *Dawkins' God: Genes, Memes and the Meaning of Life*, Oxford: Blackwell

McPherson, Thomas (1972), *The Argument from Design*, London: Macmillan

Mackie, J.L. (1955), 'Evil and Omnipotence', *Mind*, **64**, 200–12

Mackie, J.L. [1955] (1971), 'Evil and Omnipotence', in Mitchell, *The Philosophy of Religion* (see below), 92–104

Mackie, J.L. (1977), *Ethics: Inventing Right and Wrong*, Harmondsworth: Penguin

Mandelbaum, Maurice (1958), 'Darwin's Religious Views', *Journal of the History of Ideas*, **19**.3, 363–78

Mannison, Don, Michael McRobbie and Richard Routley (eds) (1980), *Environmental Philosophy*, Canberra: Department of Philosophy, Australian National University

Manson, Neil A. (ed.) (2003), *God and Design: The Teleological Argument and Modern Science*, London: Routledge

Martin, Michael (1990), *Atheism: A Philosophical Justification*, Philadelphia, PA: Temple University Press

Marx, Karl [1867–94] (1967), *Capital* (3 vols), New York: International

Masri, Al-Hafiz B.A. (1992), 'Islam and Ecology', in Khalid and O'Brien (eds), *Islam and Ecology* , 1–23

Maynard Smith, John (1981), 'Did Darwin Get it Right?', *London Review of Books*, **3**.11, 10–11

Maynard Smith, John (1982), *Evolution and the Theory of Games*, Cambridge: Cambridge University Press

Mellor, D.H. (2005), 'Accepting the Universe', *Think*, **11**, 55–64

Midgley, Mary (1979), 'Gene-Juggling', *Philosophy*, **54**, 438–58

Midgley, Mary (1985), *Evolution as a Religion: Strange Hopes and Stranger Fears*, London and New York: Methuen

Midgley, Mary (2000), 'Why Memes?', in Rose and Rose, *Alas, Poor Darwin* (see below), 67–84

Midgley, Mary (2005), Letter to *The Guardian*, 6 September

Miller, Alexander (1998), *Philosophy of Language*, London: UCL Press

Miller, Alexander (2002), 'What is the Manifestation Argument?', *Pacific Philosophical Quarterly*, **83**, 352–83

Miller, Alexander (forthcoming), 'Realism and Anti-Realism', in Lepore and Smith (eds), *Oxford Handbook on the Philosophy of Language* (see above)

Miller, Kenneth (1999), *Finding Darwin's God*, New York: HarperCollins

Miller, Kenneth (2004), 'The Flagellum Unspun: The Collapse of Irreducible Complexity', in Dembski and Ruse, *Debating Design* (see above), 81–97

Misak, C.J. (1995), *Verificationism: Its History and Prospects*, London and New York: Routledge

Mitchell, Basil (1955), 'Theology and Falsification, C', in Flew and MacIntyre, *New Essays* (see above), 103–05

Mitchell, Basil (ed.) (1971), *The Philosophy of Religion*, Oxford: Oxford University Press

Moltmann, Jürgen (1985), *God in Creation*, London, SCM

Moore, A.L. (1889), *Science and Faith*, London: Kegan Paul, Trench and Co.

Moore, A.L. (1891), 'The Christian Doctrine of God', in Gore (ed.), *Lux Mundi* (see above), 41–81

Moore, Kathleen Dean (2005), 'The Truth of the Barnacles: Rachel Carson and the Moral Significance of Wonder', *Environmental Ethics*, **27**.3, 265–77

Newton-Smith, W.H. (1981), *The Rationality of Science*, London: Routledge and Kegan Paul

Nielsen, Kai (1971), *Contemporary Critiques of Religion*, New York: Herder and Herder

Nielsen, Kai (1982), *An Introduction to the Philosophy of Religion*, London: Macmillan

Nisbet, Robert (1980), *History of the Idea of Progress*, London: Heinemann

Norris, Christopher (2004), 'Staying for an Answer', *Philosophy and Social Criticism*, **30**.7: 777–98

Norton, Bryan G. (1986), 'Conservation and Preservation: A Conceptual Rehabilitation', *Environmental Ethics*, **8**.3, 194–220

Norton, Bryan G. (1992), 'Epistemology and Intrinsic Values', *The Monist*, **75**, 208–26

Nozick, Robert (1974), *Anarchy, State and Utopia*, Oxford: Blackwell and New York: Basic Books

Nozick, Robert (1981), *Philosophical Explanations*, Oxford: Clarendon Press

Open University Course Team for Science and Belief: From Copernicus to Darwin (1974), *The New Outlook for Science*, Milton Keynes: Open University Press

O'Riordan, Tim and James Cameron (eds) (1994), *Interpreting the Precautionary Principle*, London: Cameron and May

Owen, H.P. (1971), *Concepts of Deity*, London: Macmillan

Page, Ruth (2004), 'Panentheism and Pansyntheism: God in Relation', in Clayton and Peacocke, *In Whom We Live and Move and Have Our Being* (see above), 222–32

Paley, William [1802] (1825), *Natural Theology, or Evidences of the Existence and Attributes of the Deity Collected from the Appearances of Nature*, in William Paley (ed.), *The Complete Works of William Paley, D.D.*, London: George Cowie, vol. III

Palmer, C. (1992), 'Stewardship: A Case Study in Environmental Ethics', in Ball, Goodhall, Palmer and Reader, *The Earth Beneath* (see above), 67–86

Palmer, Humphrey (1973), *Analogy: A Study of Qualification and Argument in Theology*, London: Macmillan

Parfit, Derek (1984), *Reasons and Persons*, Oxford: Clarendon Press

Passmore, John (1970), *The Perfectibility of Man*, London: Duckworth

Passmore, John (1974), *Man's Responsibility for Nature*, London: Duckworth

Payne, Peter (1996), *Design in the Universe: The Design Argument from the Perspective of Natural Law*, Claremont, CA: Claremont Graduate School

Peacocke, Arthur (1979), *Creation and the World of Science: The Bampton Lectures, 1978*, Oxford: Clarendon Press

Peacocke, Arthur (1985), 'Biological Evolution and Christian Theology – Yesterday and Today', in John Durant (ed.), *Darwinism and Divinity* (see above), 101–30

Peacocke, Arthur (1990), *Theology for a Scientific Age: Being and Becoming – Natural and Divine*, Oxford: Blackwell

Peacocke, Arthur (2004), 'Articulating God's Presence in and to the World Unveiled by the Sciences', in Clayton and Peacocke, *In Whom We Live and Move and Have Our Being* (see above), 137–54

Peirce, C.S. (1908), 'A Neglected Argument for the Reality of God', *The Hibbert Journal*, **7**, 90–112

Peirce, C.S. (1935), *Collected Papers of Charles Sanders Peirce*, (8 vols; vols 1–6, ed. C. Hartshorne and P. Weiss), Cambridge, MA: Harvard University Press

Peirce, C.S. (1958), 'Notes on Positivism', in Wiener (ed.), *Charles S. Peirce: Selected Writings* (see below), 137–41

Peirce, C.S. (1998), 'A Neglected Argument for the Reality of God', in Peirce Edition Project (ed.), *The Essential Peirce* (see below), vol. 2, 434–50

Peirce Edition Project (ed.) (1998), *The Essential Peirce: Selected Philosophical Writings* (2 vols), Bloomington and Indianapolis, IN: Indiana University Press

Penelhum, Terence (1960), 'Divine Necessity', *Mind*, **69**, 175–86

Phillips, D.Z. (1965), *The Concept of Prayer*, London: Routledge and Kegan Paul

Phillips, D.Z. (1970), *Faith and Philosophical Enquiry*, London: Routledge and Kegan Paul

Phillips, D.Z. (1993), 'On Really Believing', in Runzo, *Is God Real?* (see below), 85–107

Phillips, D.Z. (2001), *Religion and the Hermeneutics of Contemplation*, Cambridge: Cambridge University Press

Pike, Nelson (1970), *God and Timelessness*, London: Routledge and Kegan Paul

Pinker, Steven (1994), *The Language Instinct: The New Science of Language and Mind*, London: Penguin

Plantinga, Alvin (1965), 'The Free Will Defence', in M. Black (ed.), *Philosophy in America*, London: Allen and Unwin, 204–20

Plantinga, Alvin (1971), 'The Free Will Defence', in Mitchell, *The Philosophy of Religion* (see above), 105–20

Plantinga, Alvin (1980), *Does God Have a Nature?*, Milwaukee, WI: Marquette University Press

Plantinga, Alvin (1991), 'When Faith and Reason Clash: Evolution and the Bible', *Christian Scholar's Review*, **21**.1, 8–32, reprinted in Hull and Ruse, *The Philosophy of Biology* (see above), 674–97

Plantinga, Alvin (1993), *Warrant and Proper Function*, New York: Oxford University Press

Plantinga, Alvin (1997), 'Methodological Naturalism', *Perspectives on Science and Christian Faith*, **49**.3, 143–54

Plato (1929), *Timaeus*, in Plato, *Timaeus, Critias, Cleitophon, Menexenus, Epistles*, trans. R.G. Bury (Loeb Classical Library), London: Heinemann and Cambridge, MA: Harvard University Press

Popper, Karl (1959), *Logic of Scientific Discovery*, London: Hutchinson

Popper, Karl (1963), *Conjectures and Refutations*, London: Routledge and Kegan Paul

Popper, Karl (1972), *Objective Knowledge: An Evolutionary Approach*, Oxford: Clarendon Press

Putnam, Hilary (2002), *The Collapse of the Fact/Value Dichotomy and Other Essays*, Cambridge, MA: Harvard University Press

Quine, W.V.O. (1953), 'Two Dogmas of Empiricism', in W.V.O. Quine, *From a Logical Point of View: Nine Logico-Philosophical Essays*, Cambridge, MA: Harvard University Press, 20–46

Raven, Charles (1953), *Natural Religion and Christian Theology: Experience and Interpretation*, Cambridge: Cambridge University Press

Rawls, John (1972), *A Theory of Justice*, London and New York: Oxford University Press

Rea, Michael C. (2002), *World Without Design*, Oxford: Clarendon Press

Rees, Martin (2003), 'Other Universes: A Scientific Perspective', in Manson, *God and Design* (see above), 211–20

Reichenbach, Bruce (1982), *Evil and a Good God*, New York: Fordham University Press

Richards, Janet Radcliffe (2000), *Human Nature after Darwin: A Philosophical Introduction*, London and New York: Routledge

Richards, Robert J. (2002), *The Romantic Conception of Life: Science and Philosophy in the Age of Goethe*, Chicago and London: University of Chicago Press

Rolston, Holmes, III (1987), *Science and Religion: A Critical Survey*, New York: Random House

Rolston, Holmes, III (1988), *Environmental Ethics: Duties to and Values in the Natural World*, Philadelphia: Temple University Press

Rolston, Holmes, III (1992), 'Disvalues in Nature', *The Monist*, **75**, 250–78

Rolston, Holmes, III (1993), 'Biophilia, Selfish Genes, Shared Values', in Kellert and Wilson, *The Biophilia Hypothesis* (see above), 381–414

Rolston, Holmes, III (1999), *Genes, Genesis, and God: Values and their Origins in Natural and Human History*, Cambridge: Cambridge University Press

Romanes, G.J. ('Physicus') (1892), 'Supplementary Essay in Reply to a Recent Work on Theism', in G.J. Romanes, *A Candid Examination of Theism*, London: Kegan Paul, French, Trübner, 152–80

Rorty, Richard (1989), *Contingency, Irony and Solidarity*, Cambridge: Cambridge University Press

Rose, Steven (1997), *Lifelines: Biology, Freedom, Determinism*, London: Allen Lane

Rose, Hilary and Steven Rose (eds) (2000), *Alas, Poor Darwin: Arguments Against Evolutionary Psychology*, London: Jonathan Cape

Rosenthal, Abigail L. (2004), 'What Ayer Saw When He Was Dead', *Philosophy*, **79**, 507–31

Ross, J.P. (1981), *Portraying Analogy*, Cambridge: Cambridge University Press

Rousseau, Jean-Jacques [1762] (1963), *Émile*, trans. Barbara Foxley, London: Dent (Everyman's Library)

Routley, Richard (1973), 'Is There a Need for a New, an Environmental Ethic?', *Proceedings of the XVth World Congress of Philosophy*, Varna (Bulgaria), 205–10

Routley, Richard and Val Routley (1980), 'Human Chauvinism and Environmental Ethics', in Mannison, McRobbie and Routley, *Environmental Philosophy* (see above), 96–189

Rowe, William L. (1970), 'Two Criticisms of the Cosmological Argument', *The Monist*, **54**, 441–59

Rowe, William L. (1971), 'The Cosmological Argument', *Noûs*, **5**, 49–61

Runzo, Joseph (ed.) (1993), *Is God Real?*, Basingstoke: Macmillan

Ruse, Michael (1995), *Evolutionary Naturalism: Selected Essays*, London and New York: Routledge

Ruse, Michael (2001) *Can a Darwinian Be a Christian?*, Cambridge: Cambridge University Press

Ruse, Michael (2003), *Darwin and Design: Does Evolution have a Purpose?*, Cambridge, MA and London: Harvard University Press

Russell, Bertrand and F.C. Copleston (1964), 'The Existence of God: A Debate between Bertrand Russell and Father F.C. Copleston', in Hick, *The Existence of God* (see above), 167–91

Russell, C.A. (ed.) (1973), *Science and Religious Belief*, London: University of London Press

Scheffler, Israel (1996), 'Epistemology of Objectivity', in McCormick, *Starmaking* (see above), 29–58

Scherer, Donald (1983), 'Anthropocentrism, Atomism and Environmental Ethics', in Scherer and Attig, *Ethics and the Environment* (see below), 54–81

Scherer, Donald and Thomas Attig (eds) (1983), *Ethics and the Environment*, Englewood Cliffs, NJ: Prentice-Hall

Schilpp, P.A. (ed.) (1963), *The Philosophy of Rudolph Carnap*, La Salle, IL: Open Court

Schlick, Moritz (1976), 'Meaning and Verification', in Moritz Schlick, *Philosophical Papers, Volume II: 1925–1936*, eds H.L. Mulder and B.F.B. van de Velde-Schlick, Dordrecht: Reidel

Schopf, T.J.M. (ed.) (1972), *Models in Paleobiology*, San Francisco, CA: Freeman, Cooper

Shaffer, Jerome (1962), 'Existence, Predication and the Ontological Argument', *Mind*, **71**, 307–25

Singer, Peter (1981), *The Expanding Circle*, Oxford: Clarendon Press

Singer, Peter [1979] (1993), *Practical Ethics* (2nd edn), Cambridge: Cambridge University Press

Sober, Elliott (ed.) (1994), *Conceptual Issues in Evolutionary Biology* (2nd edn), Cambridge, MA: MIT Press

Sober, Elliott (1998), 'Six Sayings about Adaptationism', in Hull and Ruse, *The Philosophy of Biology* (see above), 72–86

Soulé, Michael E. (1993), 'Biophilia: Unanswered Questions', in Kellert and Wilson, *The Biophilia Hypothesis* (see above), 441–55

Stiver, Dan R. (1996), *The Philosophy of Religious Language: Sign, Symbol and Story*, Malden, MA: Blackwell

Strawson, P.F. (1959), *Individuals: An Essay in Descriptive Metaphysics*, London: Methuen

Strawson, Peter (1977), 'Scruton and Wright on Anti-Realism', *Proceedings of the Aristotelian Society*, **77**, 15–22

Swinburne, R.G. (1968), 'The Argument from Design', *Philosophy*, **43**, July, 199–212

Swinburne, R.G. (1992), *Revelation: From Metaphor to Analogy*, Oxford: Clarendon Press

Taylor, B. (ed.) (1987), *Michael Dummett: Contributions to Philosophy*, Dordrecht: Martinus Nijhoff

Taylor, Richard [1963] (1974), *Metaphysics* (2nd edn), Englewood Cliffs, NJ: Prentice-Hall

Tennant, F.R. (1928–30), *Philosophical Theology* (2 vols), Cambridge: Cambridge University Press

Throop, William (ed.) (2000), *Environmental Restoration: Ethics, Theory and Practice*; Amherst, NY: Humanity Books

Todes, Daniel Philip (1989), *Darwin Without Malthus: the Struggle for Existence in Russian Evolutionary Thought*, Oxford: Oxford University Press

Traherne, Thomas (1966), *Poems, Centuries and Three Thanksgivings*, ed. Anne Ridler, London: Oxford University Press

Trigg, Roger (1992), 'Reason and Faith – II', in Warner, *Religion and Philosophy* (see below), 33–43

Trigg, Roger (2002), *Philosophy Matters*, Malden, MA and Oxford: Blackwell

Vanhoozer, Kevin J. (1998), *Is There a Meaning in This Text?*, Leicester: Apollos

Ward, Keith (1982), *Rational Theology and the Creativity of God*, Oxford: Basil Blackwell

Ward, Keith (1996), *God, Chance and Necessity*, Oxford: Oneworld

Ward, Keith (2004), 'Theistic Evolution', in Dembski and Ruse, *Debating Design* (see above), 261–74

Ward, Keith (2004), 'The World as the Body of God: A Panentheist Metaphor', in Clayton and Peacocke, *In Whom We Live* (see above), 62–72

Warner, Martin (1992), 'Introduction', in Warner, *Religion and Philosophy* (see below), 1–21

Warner, Martin (1992), 'Language, Interpretation and Worship – I', in Warner, *Religion and Philosophy* (see below), 91–108

Warner, Martin (ed.) (1992), *Religion and Philosophy*, Cambridge: Cambridge University Press

Weil, Simone (1987), 'Are We Struggling for Justice?', trans. Marina Barabas, *Philosophical Investigations*, **10**.1, 1–10

White, John (1997), *Education and the End of Work: A New Philosophy of Work and Learning*, London: Cassell

Wiener, Philip P. (ed.) (1958), *Charles S. Peirce: Selected Writings*, New York: Dover Publications

Williams, George C. (1988), 'Huxley's Evolution and Ethics in Sociobiological Perspective', *Zygon*, **23**.4, 383–407

Williams, George C. (1996), *Plan and Purpose in Nature*, London: Weidenfeld and Nicolson

Williams, Rowan (2002), *Dimbleby Lecture*, BBC2 TV, 19 December

Wilson, Edward O. (1975), *Sociobiology: The New Synthesis*, Cambridge, MA: Belknap Press of Harvard University Press

Wilson, Edward O. (1992), *The Diversity of Life*, Cambridge, MA: Harvard University Press

Wilson, Edward O. (1993), 'Biophilia and the Conservation Ethic', in Kellert and Wilson, *The Biophilia Hypothesis* (see above), 31–41

Wollheim, Richard (ed.) (1963), *Hume on Religion*, London: Collins

Wright, Crispin (1987), *Realism, Meaning and Truth*, Oxford: Oxford University Press

Wright, Crispin (1992), *Truth and Objectivity*, Cambridge, MA: Harvard University Press

Wynn, Mark (1999), 'Natural Theology in an Ecological Mode', *Faith and Philosophy*, **16**.1, 27–42

Index

abduction 16, 23
Acquisition Challenge 32–3
adaptation 84, 97, 100–102, 105–6, 158
adaptationism 111–15, 144, 147, 160
adaptive complexity 145–6, 149–50, 160–61, 168, 204, 209
aesthetic language 13–14
agnosticism 117–18, 164
analogical use of language 25, 39
analogy
  argument from 98–9, 105, 109–10, 120, 156
  of attribution 25
  of proportionality 25–6
  theory of 25–6, 39, 45, 62, 71–2, 98, 127, 134–5
analytic propositions and statements 11, 21
animism 13
Anselm, St 78
anthropic principle 100–101
anthropocentrism 113, 129, 152–3, 159, 161, 165–6, 169, 193–4, 202
anthropogenic disruption 123–4
anthropomorphism 25–6, 62
anti-realism 1–2, 8–9, 20, 27, 29–49, 55, 62–3, 70, 89, 155
  arguments against 36–40
  in Dummett's thinking 31–6
  generic 32, 62, 206
  in Rorty's thinking 40–42
  semantic 23, 29–40
Aquinas, Thomas 73, 78, 92–3, 96, 99, 101
Aristotle and Aristotelianism 25, 207
atheism 16, 18, 41, 118, 144, 164
Atkins, Peter 69, 190
atrocities 136–9, 186, 208
attribution, analogy of 25
Augustine, St 78
'Augustinian science' 118
Austin, J.L. 42
Ayer, A.J. 9–13, 16–25, 29, 39, 52, 60, 206
Baden-Powell, R.S. 115
Barbour, Ian 24–5
Barth, Karl 69, 72, 168
Bartholomew, D.J. 150, 152
Bayesian theory 136
Behe, Michael 82, 99
Berkeley, George 40
Berry, R.J. 115
Big Bang theory 76, 91, 106, 129
biodiversity 160, 194, 204
biophilia 204–5, 210
Blank-Paper Darwinism 114, 165, 190
block-universe theory of time 78
Boethius 78–9
Braithwaite, R.B. 13, 31, 46–7
Broad, C.D. 18
Brooke, John Hedley 84, 87
Brown, Andrew 114
Brown, Patterson 73
Brown, Peter G. 194
Byrne, Peter 30, 41, 48, 206

Calvin, John 193–4
cancer 141
Carnap, Rudolph 9, 31
category mistakes 91, 102
Chomsky, Noam 179
Christianity 44, 81–2, 118–19, 122, 167, 203
Clarke, Samuel 90, 93, 96
Clayton, Philip 169–70
communication of meaning 27, 177–8, 187
composition, fallacy of 90–91
Comte, Auguste 31, 87, 163
conceptual relativism 41–2
Condorcet, Marquis de 163
conservation of natural resources 192–4
contributory value 154
Conway, David 31
Cooper, David 184
Copan, Paul 24

correspondence theory of truth 31, 40, 42
Cosmological Argument 48–9, 58, 63, 67, 72–5, 89–99, 103, 108, 115–16, 119–21, 161, 207
Cottingham, John 186–91, 210
counterfactuals 11, 146
Craig, William Lane 24, 78, 105–7, 169
'cranes' (Dennett) 113–14, 179, 185
creation
  belief in 1, 24, 27, 40–41, 49, 56, 61, 63, 71, 73, 76, 166, 203–10
  and falsification 56–9
  operating through natural processes 166–7
  as renunciation 71
  and time 76–80
  without creationism 80–83
creationism 1, 12, 43–4, 63, 80–83, 99–100, 114, 119–20, 147, 149, 207
critical realism 49, 73, 164, 206
culture, evolution of 184, 185
Cupitt, Don 30, 42–6, 55, 62, 206

Darwin, Charles 81–7, 89, 97–8, 104, 109–11, 116, 121–2, 125, 130, 139, 163, 166, 207–8
'Darwin wars' 114
Darwinism 1–3, 12, 80–83, 91, 97, 101–2, 108–24, 130, 144–50, 163–5, 170–71, 179, 181, 185, 201, 207–9
  generic 112, 148–9
  and meaningful life 189–91
  and theism 115–22, 126, 130, 142–50, 168
Darwinism-plus 150
Davies, Brian 79
Davis, Stephen 45, 93, 102–3, 106, 133–8
Dawkins, Richard 97, 103, 110–16, 119, 139, 148–50, 152, 189–90, 199, 204, 208
  on predation 143–7
deism 83, 167, 207
Dembski, William A. 82
Dennett, Daniel 112–13, 116, 160, 168, 175–81, 194, 204, 209
  Tower of Generate-and-Test 179–86, 191
Derrida, Jacques 17
Design Argument 85–6, 89–90, 97–108, 115–16, 119–21, 207
  Hume's critique of 101–6
Devitt, Michael 30
disvalues in nature 121, 128–30
Dobzhansky, Theodosius 83, 118
Dostoevsky, Fyodor 137
Drees, Willem B. 199
Duhem, Pierre 52–5
Dummett, Michael 7, 10, 17, 22–3, 29–36, 53, 62, 206
Dupré, John 89–90, 96–8, 101, 112, 114–19
Durrant, Michael 76–8

Einstein, Albert 51
Eldredge, Niles 111
emotivism 30
Empedocles 51, 54, 63
empiricism 9, 31
Enlightenment thinking 41, 97, 163
Epicurus 122
epistemes 80–81
equivocation 25–6, 62
error theory 29–30
ethical discourse 13–14, 29–30, 116, 155–7
ethical responsibilities 197–200, 209
Evanoff, Richard J. 198–9
Evans, J.L. 19, 21
evil
  avoidable 122
  problem of 55–6, 104, 130, 133–6
  see also moral evil; natural evil
evolution, progressivism in 162–5, 166, 180–81, 186, 209
evolutionary naturalism 163
evolutionism 1, 145; *see also* theistic evolutionism
extinction of species 37, 122, 142
extrinsic value 154, 176

fact/value distinction 116, 155
falsifiability 2, 9, 20, 51–6, 118, 205–7
  and creation 56–9
  of religious statements and explanations 59–61
Farrer, A.M. 117, 167
Feigl, Herbert 22

Feinberg, Joel 193
Fermat's last theorem 37
Feuerbach, Ludwig 47–9
Findlay, J.N. 74
Fine Tuning Argument 92, 100–101, 104–6, 117, 120–23, 153, 159, 207–8
Fisher, Ronald 83, 110, 118, 147, 160–61
Flew, Antony 24–7, 51–63, 203, 206–7
flourishing of living creatures 142–3, 158–62, 169, 189–90, 194, 198
Foot, Philippa 156
Ford, E.B. 160–61
Foster, Michael 24
Foucault, Michel 80–81
Free Will Defence 133–8, 186, 208
Freud, Sigmund 14

Gallagher, Kenneth T. 110
Geach, Peter 77–8
Gene-Machine Darwinism 152, 165, 175, 179, 190–91, 195, 210
Genesis, Book of 72, 81, 129
genetics 83, 110–15
Gillespie, Neal C. 80–82
global warming 124, 199
God
  answerability to 27, 49, 202–3
  attributes of 25, 47–8, 55–7, 67–8, 71–2, 77–9, 96, 120–21, 133–5
  existence of 16, 18, 26, 46–7, 54, 74, 89–108, 115, 120–50, 151–62, 205
  supposed necessity of 73–6, 96
Goldbach's Conjecture 54
Goodpaster, Kenneth 157, 198
Gould, Stephen Jay 110–18, 122, 125, 147, 149, 159, 168, 179, 184, 190, 208
Gray, Asa 81, 121
'Gregorian creatures' 181–5
Gregory, Richard 180

Hale, Sir Matthew 192–3
Hannam, James 69
Hardy, Thomas 128, 137
Hare, R.M. 31–4, 46, 55
Hartshorne, Charles 169
Hebblethwaite, Brian 45, 128, 137
Hegel, G.W.F. 43, 163
Heidegger, Martin 17
Hepburn, R.W. 89
Hick, John 18–19, 47, 59–60, 137–8
historical language 14
historical progress 163
Hook, Bill 175–8, 187, 191, 194
Huiser, Pieter J. 70
Hull, David 163–4
Hume, David 9–10, 12, 30–31, 75, 87, 94–8, 115–16, 119–25, 130, 138–40, 207–8
  critique of Design Argument 101–6
Hurlbutt, Robert H. 99
Huxley, Thomas Henry 111, 116, 126
hypothesis-testing 52, 180, 181, 184

Ichneumonidae 121–2, 125, 129–30, 139
idealism 40, 45
immanence 69, 167–70, 209
imperfection, argument from 71
infinite sets, explanation of 95
inherent value 154
instrumental value 153–4
intelligence, development of 178, 181, 185, 194
Intelligent Design 1–2, 82, 100, 120
intrinsic value 2, 143, 151–71, 175–6, 186, 193–4, 198–200, 209
Irenaeus and Irenaean theodicy 137–8
irreducible complexity 1, 82, 99–101, 105, 120
Islam 202–3

Jonas, Hans 197–200
Judaism 203

Kant, Immanuel 43, 45, 48, 75, 102, 163
Kenny, Anthony 78–9, 134
Kerr, Fergus 46
Kimura, Motoo 112
Kingsley, Charles 86, 167
Krebs cycle 100
Kripke, Saul 37
Kropotkin, Petr 126

Lamarckianism 144
language, use of 29, 180–83, 184; *see also* religious language
Le Poidevin, Robin 74, 101, 107–8

Leibniz, Gottfried Wilhelm 15, 17, 23, 57, 93, 96
Lewis, H.D. 69–70
Lewontin, Richard 111–14, 147–8, 208
libertarian freedom 133–6, 152–3, 185–6
logical positivism 10–22, 31–2
Lucas, J.R. 78
Lucretius 15, 107
Luther, Martin 187
Lyell, Charles 85

McDowell, John 37
McGrath, Alister 115, 144
Mach, Ernst 9
Mackie, J.L. 29–30, 136, 155
MacKinnon, Donald 24–6, 203
McTaggart, J.E. 18
Malthus, Thomas Robert 84, 109
Mandelbaum, Maurice 84–6
Manifestation Challenge 33–4
Marx, Karl 14, 43, 163, 202
Marxism 51, 54
materialism 114, 118, 165
Matter First Darwinism 114, 116, 190–91
Maynard Smith, John 111
meaning
  holistic view of 39, 52–3
  in language and in life 1, 7–8, 178, 194, 209
  and mind 177, 179
  theory of 1–2, 7–13, 19–23, 26–30, 31–42, 49, 62
  and value 175–6
meaningful action 175–8, 184–9, 194–5
meaningful life 177–8, 186–95, 205, 209–10
  and Darwinism 189–91
meaningful work 187–9
Mellor, D.H. 107–8
memes 2–3, 112–13, 164, 179, 183–4, 190
Mendel, Gregor 110, 163
metaphorical language 39
metaphysics 10, 12, 15–18, 29, 40–49, 56–8, 67–80, 89–108, 133–71
  of stewardship 200–203, 209–10
methodological naturalism 83, 117–18
Midgley, Mary 112–13, 166, 181–2, 189–91
migrations 127
Miller, Alexander 34–8
Mind First Darwinism 3, 113–14, 165, 190, 210
minimalism 67, 69
miracles 118, 124–5, 138–9, 147, 208
Misak, C.J. 12, 15, 16, 23, 52, 61
Mitchell, Basil 55–6, 63
Moltmann, Jürgen 168–71
Monod, Jacques 159
Moore, Aubrey 167–70
moral evil 133–4, 136–9, 150, 171, 186, 208
moral judgements 156, 198–200
mutation and mutationism 144, 148, 159–60, 165
myths 51, 155

natural evil 86–7, 109, 122–3, 125, 130–31, 133, 139–50, 171, 208
natural laws 76, 83, 99–108, 115, 123–5, 140–41, 145–8, 150, 165, 207
natural selection 84–7, 89, 91, 95, 97, 100, 105, 109–25, 129–30, 142–50, 152, 162–8, 176–83, 190, 207–8
natural theology 58, 89–108, 151–62, 167
near-to-death 17–18, 206
neo-Darwinism 191
neo-Platonism 75
Neurath, Otto 31
New Creationism 82
Newman, John Henry 115, 117
Newton, Isaac 98–101
Nietzsche, Friedrich 31, 43
NOMA (non-overlapping magisteria) principle 116–17, 190
normativity of meaning 36–8
Norris, Christopher 35
Nozick, Robert 57–8, 93, 156, 158

objectivity of meaning 37–8
Ogden, Schubert 169
omnipotence 79, 134–5
omniscience 79, 134
Ontological Argument 74–5, 90, 96
Owen, H.P. 69–73, 169–70

pain 87, 104, 122–3, 139–42, 158

Paley, William 85, 98, 101, 110, 115, 119–20, 144, 179
Palmer, Humphrey 71
panentheism 67, 69–70, 168–71, 209
Pangaea 96
pantheism 13, 67–70, 75, 168–9
parasitism 84, 121–2, 126–30, 139–40, 143–8, 208
Parfit, Derek 156
Parmenides 51, 54, 63
Paul, St 168
Payne, Peter 102
Peacocke, Arthur 49, 149, 168–71
Peirce, C.S. 9, 15–16, 21, 23, 43, 206
Penelhum, Terence 74, 90
perspectivism 43, 45, 157
Phillips, D.Z. 45–9, 62, 70–71, 89–91, 206
philosophy, role of 7, 9, 181
philosophy of religion 30, 46, 49, 206
philosophy of science 30, 44, 51, 60, 105, 117, 163–4
physical theology 120
Pike, Nelson 78
Plantinga, Alvin 82, 114–19, 136, 164
Plato 17, 67, 71, 76, 82
Platonism 44, 169
Pluto 96, 101
Popper, Karl 12, 51–4, 60, 63, 180
'Popperian creatures' 181–5
positivism 9, 22–3, 80–81, 155, 206; *see also* logical positivism
pragmatism 9, 13, 15, 21, 23, 31, 43
Precautionary Principle 199
predation 84, 128–30, 134, 139, 142–8, 162, 182, 208
  Dawkins on 143–7
prescriptivism 30
primary goods 153
probability, concepts of 106–8, 208
process theology 77, 82, 168–9
proportionality, analogy of 25–6
psycho-analysis 14, 51
psychology, evolutionary 114, 175–85
'push-up, lock-up' effect 149–50
pyramid of design 179

quality of life 162, 166
quantifier shift fallacy 91
Quine, W.V.O. 11, 20, 39, 52–5, 59, 206

randomness 121, 150, 152, 159, 165
Raven, Charles 161, 168
Rawls, John 153
Rea, Michael C. 100–101, 108
realism 29–49, 55, 62, 73, 203, 206
  generic 32–40
  see also semantic realism
reciprocal altruism 199
Recognition Challenge 35–6
Rees, Martin 106–7
refutability of a theory *see* falsifiability
regularities of succession, argument from 99, 101, 102–6, 207–8
Reichenbach, Bruce 79, 136, 140–41, 146
religious language 12–13, 24, 26, 45, 47, 49, 59–61, 206–7
  implications of verificationism for 13–16
revelations 59, 63, 72
Richards, Janet Radcliffe 113–14, 118, 179, 190
Richards, Robert J. 84, 85
Rolston, Holmes 127–30, 139–50, 159–60, 204
Romanes, George 128
Rorty, Richard 40–43, 62, 206
Rosenthal, Abigail L. 9, 17–18
Rousseau, Jean-Jacques 96, 167
Routley, Richard 156–7
Rowe, William 48, 90–95
Ruse, Michael 81–3, 100, 103, 109–12, 116–19, 144–52, 156, 159–64, 168, 204, 208–9
Russell, Bertrand 15–16, 91, 93
Ryle, Gilbert 91

Saint-Simon, Comte de 163
saltationism 110–11
Scherer, Donald 148, 157
Schleiermacher, Friedrich 78
Schlick, Moritz 9, 61, 188
scientific advance 163–4, 184, 191
scientific method 10, 51–2, 61, 80, 83, 118–19
  and religion 116–17, 118–19
second-order beliefs 180–83
self, the, concept of 182–3

self-awareness 187–9, 191
semantic realism 34–5, 38
semantic theories 7–8, 17, 23, 29, 54, 62
Shaffer, Jerome 74
Shakespeare, William 27
Singer, Peter 199–200
'Skinnerian creatures' 179–82, 185
'skyhooks' (Dennett) 113, 179, 185
Smith, Quentin 30
sociobiology 199–200
solar system 98–100
Spencer, Herbert 163
Spinoza, Baruch 17, 75
stewardship of the natural world 2–3, 27, 49, 175, 191–206
  metaphysics of 200–203, 209–10
Stoicism 68
Strawson, Peter 34, 36
subjectivism 157
suffering 104, 122, 139–48, 159, 190
Sufficient Reason, Principle of 57–63, 90, 93–6, 103–8, 120
supernaturalism 92
Swinburne, R.G. 30, 101–3
synthetic propositions and statements 11, 21
systemic value 127, 160

Taylor, Richard 93
Teleological Argument 89–90, 97–108, 109, 130, 151–71
Tennant, F.R. 141
theism 12, 23–4, 27, 30, 41, 53–63, 67–91, 103, 144–5, 151–71, 190–92
  compatibility with evil 133–6
  and Darwinism 115–21, 130, 168
  and stewardship 201–10
theistic evolutionism 164–6, 209
theodicy 56, 121–5, 128–30, 133–50, 207
Think (journal) 101
timelessness of God 76–80, 207
Tindal, Matthew 167
Toland, John 167
tools, use of 180, 183
Traherne, Thomas 205
transcendence 63, 69–73, 79, 168
Trigg, Roger 41
'truth-bearers' and 'truth-makers' (Byrne) 41
Truth-Conditional Conception of Meaning and Understanding (TCCMU) 38–40
truth-conditions 32–6, 38–40, 206
Twain, Mark 125

ultra-Darwinism 112, 118–19, 146–7
uniformitarianism 85
United Nations 202
utterances, meaning and understanding of 19–20, 39, 186

value, arguments from 143, 151–71, 209
verificationism 8–18, 27–8, 31, 52, 54, 61, 62, 155
  implications for religious language 13–16
  objections to 19–24, 62
Vienna Circle 9, 52
Voltaire 122, 167
Vrba, Elisabeth S. 111–12

Waal, Frans de 126–7
Ward, Keith 2, 72, 75, 99, 104–5, 119, 144–66, 169–71, 209
Warner, Martin 29–30, 43
waste, problem of 141–2, 146, 208
Waterland, Daniel 96
Weil, Simone 71
White, John 186–8
Wiles, Andrew 37
Williams, George C. 126–7
Wilson, Edward O. 129, 160, 191, 204–5, 210
Wisdom, John 53
Wittgenstein, Ludwig 31, 36–7, 43–7
Wright, Crispin 32–8, 62, 206